中国人类学

第二辑

周永明 / 主编

图书在版编目(CIP)数据

中国人类学. 第2辑/周永明主编. —北京:商务印书馆,2020
ISBN 978-7-100-19103-6

Ⅰ.①中… Ⅱ.①周… Ⅲ.①人类学—研究
Ⅳ.①Q98

中国版本图书馆CIP数据核字(2020)第181470号

中国人类学
第二辑
周永明 主编

商 务 印 书 馆 出 版
(北京王府井大街36号 邮政编码100710)
商 务 印 书 馆 发 行
北京艺辉伊航图文有限公司印刷
ISBN 978-7-100-19103-6

2020年11月第1版 开本787×1092 1/16
2020年11月北京第1次印刷 印张17¼
定价:88.00元

《中国人类学》编辑部地址：

南方科技大学社会科学中心暨社会科学高等研究院
广东省深圳市南山区学苑大道 1088 号
邮　　编：518055
电子邮箱：caeditor@163. com

目　录

医学人类学专辑

学人访谈

理论与实践

青年新作

书 评

编后记

医学人类学专辑

本辑特约编辑

冯珠娣（Judith Farquhar）

赖立里

前　言

赖立里　冯珠娣*

本期《中国人类学》为“医学人类学特辑”，在此有必要对相关概念做一些解释。过去几十年，随着人类学这一学科在中国的学术体系中逐渐发展起来，在中国开展研究的社会人类学家们已经将健康和医疗纳入视野，并且做了许多重要的工作。但是直到近些年，“医学人类学”才作为一个专业的研究领域，开始步入正轨。2016 年，中国人类学民族学研究会（CUAES）成立了医学人类学专业委员会，中国的医学人类学研究者至此有了一个官方组织。

最近几年，本辑编者之一冯珠娣（Judith Farquhar) 多次受邀到中国的各大高校，面向人类学与社会学专业的学生讲授医学人类学。北美医学人类学家凯博文（Arthur Kleinman）与玛格丽特·洛克（Margaret Lock）也到访中国，介绍了他们的研究成果。本辑的另一位编者赖立里，在北京大学颇具规模的医学人文学院中讲授人类学、科学技术与社会，她的研究注重打通医学史、医药和文献研究以及健康社会学的界限。还有一些值得关注的研究者，例如：张文义一直在中

* 赖立里，北京大学医学人文学院。
冯珠娣（Judith Farquhar），芝加哥大学人类学系。

山大学讲授医学人类学的相关课程（本辑收录了他对冯珠娣的一个访谈）。景军在清华大学培养了一批新生代的医学人类学青年学者（其中包括余成普博士，现任教于中山大学，他同时也是医学人类学专业委员会的秘书长）。社会人类学家邵京从经济人类学的角度对1990年代因卖血而感染HIV/患艾滋病的社群进行了深入研究。另外还有本辑文章作者之一翁乃群，长期致力于探索少数民族文化习俗对健康和医疗服务的影响。

医学人类学并非专攻一门的学问，而是一个多元化的领域。我辈学者在探究医疗的社会史、健康与疾病的诸面相时，殊途同归地认识到，历史上存在着远远不止一种“医学”。事实上，我们在编辑时，正是以凸显医学议题的多样性为希冀，发现、发表中国这方面的一些研究，包括关于身体与疾病的知识、各种治疗实践、健康差异化、社会具身化、患病经验和患者叙事、生物科学与技术、传统民间医疗体系、就医机构的形式转变、现代疾病的定义与分布，以及学术史等，不一而足，难以尽述。仅看这冰山一角，我们就不得不承认，即使在今时今日，医学也不是“只此一家”，它并非特指某一领域或某一专业，“医学”一词本身就像“科学”一样，含义广博而又莫衷一是。众所周知，有关科学本质的争论从未停止。

有一种定义科学的方式，宣称“科学”即科学家之所为。或许最好也这样去理解医学人类学，它就是我们这些医学人类学家之所为。这种行为主义的界定方式诚然称不上精确，然而，如果我们承认存在着多种多样的医学，那么我们也必须承认，存在着多种多样的医学人类学。

本期《中国人类学》所选的这一系列文章极好地说明了这一点。翁乃群教授将纳日族群中出现的一个社会健康问题——“酗酒”——置于当地的文化语境中加以考察，作为一名民族志写作者，他在纳日族群中做了多年的田野调查，他的文章是使用经典人类学方法开展研究的范例。他细致地描述了近些年纳日社区中酒的生产与消费方式的变化，深刻地认识到过度饮酒这一行为如何在纳日人的礼俗中得到解释；同时还指出了家屋在纳日村寨中的重要地位。他将一个简单化的（成问题的）医学类别拆解开来，显示出对社会健康的异变做医学化处理——给大量饮酒行为贴上“酗酒”的标签——可能是颇具误导性的，他身体力行的长期民族志实践比任何理论都更有力。他的研究既在公众健康层面富有意义，也有助于让更多人了解纳日族群。

张艳华关注的是一股更为城市化的健康潮流，她与翁乃群教授的共通之处在于，都是从一个认知度很高的健康问题进入更广阔的文化内涵和社会意义层面，寻求在医学之外理解人的具身化（embodiment）。她注意到近来在现代城市中，对身体进行“排毒”成为一种时尚，张教授反思了这种感到需要被清洁并付诸行动的身体。它或已不是那种在艰苦的历史时期被塑造与扭曲的中国人的身体，那种身体可能因不再承受集体主义对欲望的克制和剥夺而渴望得到更好的滋养（中医术语叫“补”，她在此特别提到冯珠娣《饕餮之欲》一书中的观点）。如今这种需要排毒的身体害怕脂肪沉积、营养过剩导致消化不良、环境污染的侵害、白领生活习惯养成的惰性。尽管张教授与翁教授的“田野”非常不同，但两位作者都引导我们透过一个或多或少是由生物学所定义的医学现象，看到不仅属于“本土生物学”（local biologies，由洛克提出，翁乃群在文中引用）的复杂经验。他们的民族志所呈现的健康异变，既有文化特性，又有时代特性。

马志莹的文章是另一种人类学的演绎。她思考的是重度精神疾病、机构收治、2013 年起实施的《精神卫生法》和家庭成员试图“管”病人时面临的困境之间的纠葛。她的研究不是质疑一个疾病类别，也非透过疾病看社会，而是关注医学与国家体制如何管理那些难以管理的长期病患。她谨慎地对一些制约着所有相关行动者的新现结构进行了扎实的社会学考察，她的焦点仅对准国家精神病治疗机构，并无其他的政治指涉。活在当代中国，阅读她的文章，不难理解她所讨论的“管”的模棱含义——既可理解成关爱，也可理解成强硬的控制，同时不难想到，存在这种家长式作风倾向的情况又何止于此。必须对棘手经验进行管理的自上而下的压力，带来了各种问题，借用马志莹的含蓄说法，这些问题让我们很多人深深感到一种集体性的“伦理不安”。

一些人类学家认为，民族志和历史没有太大区别，所有这方面的论著都致力于把握他们所描述的特定社会形态的历史性。在本辑中，“做历史”的人类学家艾理克（Eric Karchmer）在传统中医药的世界中发现了一种当代意识形态构成，对此进行了系谱学研究。中医仅擅长治慢性病，而且“见效慢”，艾理克追溯了这种观念是如何发展起来的。他采访了一批中国最资深的老中医，积累了丰富的访谈资料，在此基础上以人类学方法挑战了中医历史与社会研究中很多时候被认为是约定俗成的东西。他揭示出，其实是直到“西方的”生物医学在中国（乃至世界）占据了认识论上的霸权地位之后，中医的大夫、病人和评论者才开始认为这一博大精深的东亚思想脉络与治疗体系是“慢”的，原本有相当多的医治急性病症见效迅速的中医知识被边缘化了。艾理克的长期研究项目是探究现代形式的中医在何种意义上可以说具有后殖民主义与东方主义的性质，这篇史述与该项目一脉相承。

何伟亚（James Hevia）对一种骆驼疾病的历史考察在本辑中可能显得有些与众不同，但正如他过去对在华西方帝国主义的研究一样，在很多方面让人类学家感到相通相惜。何伟亚紧密结合相关的科学与技术研究，同时，重视动态网络中非人类的“行动者”，这是人类学一个新的着力之处。他之前研究过英国在印度实施的殖民政策和军事管理所造成的诸多后果，在此，他又做了一个延续性的探讨。对骆驼苏拉病的研究将很多看似风马牛不相及的事情联系到了一起：各种殖民“改良”方案、旁遮普地区的灌溉、英属印度在阿富汗的军事冒险、田野中的实验室和非洲的牛、当地的骆驼行家（sarwans）和军事后勤补给计划、吸血蝇和逐渐消失的荒地……与马志莹在广州对专家、政府和普通家庭之间的张力关系的调查异曲同工，何伟亚的研究通过一种如今已被充分认清的疾病，揭示了一个远非完美的统治状态下的种种龃龉。

最后，我们要感谢满怀热忱地在中国探索医学人类学的张文义，他采访冯珠娣的内容收录在本辑中。我们特别建议读者留意他在访谈中所提的问题，从中可以窥见这个激动人心的领域的未来。

社会文化与健康

——田野志研究视角下“健康中国”建设之思考

翁乃群*

我们这里讨论的“健康”应该是源自英文“health”的概念。根据维基百科的定义，“health”是指生物有机体的机能或代谢效率水平。对于人类来说，“health”是指在面对身体、精神和社会挑战时，个体或社群的适应和自理能力。1948年《世界卫生组织宪章》序言对“health”作了更为宽泛的定义。在其定义中，“health”是指“不仅无疾病或衰弱，而且身体、精神和社会生活都完全（complete）良好的状态”。世卫组织的定义引起了较多的争议，主要是由于定义中使用了“完全”一词，致使其缺乏可操作性。[1] 后来，一些学者或机构提出了各种不同的定义，其中新近提出的定义主要把健康（health）和个人满意结合在一起。[2] 除了医疗服务外，诸如个人的社会背景、生活方式、经济和社会条件、精神状态等重要社会文化因素也影响人的健康。[3] 对于人类来说，“健康”远不只是关涉生物体的机能和代谢效率水平，而且更关涉诸如政治、经济和社会文化等人类生存的社会环境问题。人

* 翁乃群，中国社会科学院民族学与人类学研究所。

类的生物体机能和新陈代谢效率不仅受到自然生理因素的制约和影响，同时也受到社会文化环境的制约和影响。因此，“健康”问题不仅是生物医学的问题，更是人文和社会科学的问题。反之，“健康”问题也对政治、经济和社会人文有着巨大的影响。

早在 19 世纪中叶，既是医生、分子病理学的创始人、现代公共卫生的先驱、人类学家，又是政治家的德国人鲁道夫·弗卓（Rudolf Virchow）就指出：“医学是一门社会科学，政治无非是更大的医学。”他还认为：“医生是穷人的天然律师，很大程度上许多社会问题应当由他们解决。”[4] 在他发表上述论断的前一年，他的同事所罗门·纽曼（Solomon Neumann）就指出：“大多数妨碍人们充分享受生活或使相当多人不能尽享天年的疾病，并非自然体格问题，而是人为造成的社会境况所致，看清这一点无需特别的证明。就其本质而言，医学是一门社会科学，在医学这一本质被真正地认识和承认之前，我们无法享受到其益处而只能是满足于一副空壳”（Trostle, 1986: 46）。与弗卓和纽曼同时代的恩格斯通过长达近两年的实地观察和对大量档案资料的研究，在《英国工人阶级状况》一文中深刻地分析了疾病，尤其是诸如肺结核、伤寒等传染病在当时曼彻斯特英国工人中蔓延的政治经济和社会文化原因。

中国的地理环境、气候条件和自然生态环境呈现出高度的多样性，在社会文化上，是复杂多元的，在社会经济发展上，城乡和地区之间的差距仍非常巨大。在法治建设和政府治理水平上，包括关涉健康的诸如食品、药品、交通和生产安全监管方面，以及环保法律、法规的贯彻执行等方面，各地仍存在明显的差异。在医疗卫生资源分配上，城乡和地区之间不仅在硬件上，更在软件上存在着严重的不平衡。这些“多元”、“多样”、“差异”、“不平衡”构成了“健康中国”建设的重要语境。换言之，我国的健康问题是不具有同质性的。因此，我们的健康问题研究不能满足于停留在宏观研究上，同时还需要有许许多多微观的“精准”研究，并且这些研究需要有多学科的参与。

一、酗酒问题的地方语境

在我国一般社会认知中，除了信仰伊斯兰教的民族之外，其余的少数民族都有酿酒、饮酒的悠久历史和深厚的酒文化底蕴。饮酒成为了少数民族社会文化的重要标签。一些民族酒具也成为了其民族的一种物质文化符号。酒不仅是一种饮料，在我国还常被当作养生饮品、治疗病痛或泡制药物的液体媒介，更是社会仪式之中、礼尚往来之际不可或缺的重要媒介或礼物，促进社会关系的凝聚和再生产。众所周知，除了上述正面作用外，饮酒过量会给人们的身体和社会带来各种严重的负面问题。根据2015年世界卫生组织发布的《2014年酒精与健康全球状况报告》（Global Status Report on Alcohol and Health 2014）的数据显示，在我国，15岁（含）以上人口中的人均饮酒的纯酒精含量，在2003年至2005年为4. 9升，而在2008年至2010年为6. 7升。其中2008至2010年的男性人均饮酒的纯酒精含量是10. 9升，而女性为2. 2升。2010年全国15岁以上曾偶尔大量饮酒（在过去30天内至少曾有一次饮酒的纯酒精含量达到60克或以上）的男性，占全国15岁以上男性人口的14. 2%，而相应年龄区间大量饮酒的女性，占比为0. 7%。大量饮酒的男女占全国该年龄区间总人口的7. 6%。在2010年15岁以上饮酒人口中，男性的人均纯酒精饮酒含量为18. 7升，女性为6. 7升。2010年15岁以上人口中在12个月内，酒精依赖和饮酒过量的男性占全国该年龄段男性人口的9. 3%，患有酒精依赖的占4. 5%。而相应的女性人口中的上述数据，则分别为0. 2%和0. 1%。全国15岁以上总人口中，酒精依赖和饮酒过量的占

4. 9%，而患有酒精依赖者为 2. 4%（西太平洋地区的这项数据分别为 4. 6% 和 2. 3%）。2012 年全国 15 岁以上的每 10 万人口中死于肝硬化的男性有 9. 9 人，女性有 5. 8 人。上述因肝硬化死亡的男女，与饮酒有相关性的百分比分别为 73% 和 59. 8%。全国每 10 万人中，因交通事故死亡的人数，男性为 30. 5，女性为 15. 6。其中与饮酒有相关性的百分比分别为 22. 2% 和 4. 4%。[5]

在 1987、1988、1989 年的晚秋和冬季，笔者曾多次深入川滇边境的凉山彝族自治州，在盐源县、木里藏族自治县和丽江市宁蒗彝族自治县的几个乡镇村落开展长时段的人类学田野调查。在相隔十年之后的 1999 年冬季到 2013 年秋季，笔者又曾先后多次深入上述地区开展田野调查。在踏入上述田野点之始，笔者就目睹和体验了该地区各族群浓浓的酒文化。不论是行进在路途上，还是住宿、做客村户，不论是正逢年节还是平常日子，不论是遇到村户举办红白喜事，还是举行祭神、拜祖、祛病除秽的仪式，都少不了要消费大量的酒。其中，每逢重大年节和红白喜事，各村户消费的酒量最甚，各村户都要准备大量的自酿美酒款待亲朋好友，供奉神灵和祖先等。在这些少数民族社区里，自酿黄酒或自烤白酒是遇逢年节和红白喜事时，亲戚家户间用于维系和再生产亲属关系的重要交换礼物。在笔者开展田野研究的一个多民族乡，村民外出，尤其是男性，不论是徒步还是骑骡马，行囊里通常都会带有自酿黄酒或从商店里买的啤酒或白酒。每每他们在途中遇到熟人，都会停下互相问候，并各自拿出所带的酒互相敬酒。[6]

由于地处山区，村民酿酒多用玉米、大麦或青稞。酒曲是自制的，用山上采来的草药在石臼里舂碎，与谷类面粉揉和成大枣粒状，发酵酶化后用细麻绳穿起挂在屋里墙上和屋梁上储存备用。村户酿制黄酒或“烤”（当地汉语方言，意为蒸馏）制白酒，以及事前制作的酒曲通常均由家中主妇一人操作。每个村户家里都会有大铁锅、多个大的陶制缩口坛、大甑子等用来酿酒或烤酒的工具和容器。遇逢自家或亲属家中将举办红白喜事，或临近节庆，各村户主妇都会忙于酿制数量较大的黄酒和烤制白酒；平常，尤其临近农忙季节，她们也会事先酿制黄酒或烤制白酒以供农忙时消费。在他们的社会声誉体系中，生活较为殷实的家户通常都备有自制黄酒和白酒，也因此获得村民的赞誉。这些家户的主妇也会因此受到村民的夸赞。村民在日常闲谈中，常会对各家主妇酿制的黄酒或烤制的白酒评头论足。实际上，这些都构成了村民评价各村户主妇主理家户经济和生活能力水平的重要指标。

值得指出的是，在这些族群传统自给自足的社会经济生活中，各村户每年酒的消费量，一方面受制于各户当年的粮食收获量，另一方面又与各村户主妇是否能干和具有较好的管理能力密切相关。也就是说，这些社区自身对酒的消费有着以家户为单位的自我控制力。而各村户酒的消费量则由其主妇所掌控。改革开放之后，上个世纪 80 年代初，上述地区农村全面实行了家庭联产承包责任制。与此同时，农村商品市场也逐步得到发展。到了 80 年代中期，商品酒，包括外来的各种瓶装酒和乡镇个体小酒坊生产的散装白酒进入了乡镇和村落的小商铺，各社区自身原有的对酒消费的控制力便逐渐被削弱。起初，一些手上握有现金或固定经济收入的人群，诸如乡镇干部、教师、乡镇卫生所医生、乡镇供销社工作人员等，成为了首先得以冲破社区原有控制力，消费商品酒的群体。直至 90 年代，随着从事非农业劳动的村民愈来愈多，家户成员中，尤其是男性成员手头握有现金的增加，他们对酒的消费也不再完全依赖于自酿黄酒或自烤白酒。社区原先传统形成的，以各家户主妇集体主掌的消费控制权便走向了末日，从而造成了社区酒消费的失序。根据调查，在这些地区乡镇和村寨的各小商铺里，酒和烟毫无例外是销售量最多的商品，销售额可占全年商品总销售额的 70% 以上。

在笔者调查的社区里，酒消费的失序最初主要发生在少数民族公职人员，即少数民族精英层中。作为社会能动主体，他们集多种社会身份或角色于一身。除了作为家户成员（诸如：儿子、女儿、丈夫、妻子、父亲、母亲、女婿、媳妇等）的身份或角色之外，作为民族的精英，他们还有公务员、公职人员、教师、党员等身份或角色。在许多场合下，他们面对本民族与主流社会的文化差异，尤其是在“人观”、宇宙观、价值观、伦理道德、声誉体系和宗教信仰等方面的差异，或是面对由自身多重身份或角色所带来的不同的社会责任，往往会深感困惑而不知所措，并在心理上产生难以排解的压力。于是饮酒便成为了他们排解心理压抑的麻醉剂。

在特定的社会风气背景下，喝酒还被许多乡镇基层干部当成应对上下级关系，以及与外面社会进行沟通、公关和建立“良好”关系，即按社会学所说积累“社会资本”的有效方式。正是在上述多种社会文化因素的作用下，笔者在田野点上遇到了数量较多的因长期饮酒而患有酒精依赖和严重健康问题的乡、县干部。自 1988 年以来，笔者就在一个人口不足二千的乡里，先后结识了患有因饮酒引起严重健康问题的13位有工资收入者，他们都是来自该乡两个村寨的村户。1988年，这两个村寨的人口分别只有205和319人。这13人中，县级干部1人、乡级干部1人、区医院医生1人（后调到县医院）、县防疫站医生1人、区中学职工1人、乡卫生院医生1人，乡小学教师3人，乡炊事员1人、乡供销社职工2人、乡小水电站工人1人。至2008年，上述13人中已有8人亡故。其中，有4人亡故时才三十来岁，2人亡故时才四十来岁，而另外2人亡故时也只有五十来岁。值得指出的是，该调查数据只是来自该乡两个村寨（3个村民小组），其中所涉人口也只是全乡人口的三分之一左右。

如果说，在集体计划经济下，外来的商品酒非常有限，该地区酒消费主要靠的是村户自制，社区对酒的消费具有一定的自我控制力，那么改革开放后，随着商品酒的进入，以及乡、镇内小型个体家庭酒坊的重新兴起，社区逐步丧失了对酒消费的自我控制力。在这些偏远地区食品卫生监督机制的完全缺失，使市场假酒和“毒酒”的流通毫无障碍。加上前面所述的社会文化差异给部分少数民族精英带来的心理困惑和压抑，便构成了他们酗酒、酒精依赖，损害健康，直至导致早逝的重要背景。虽然，这些精英在与笔者聊天时，往往会提及市场上不乏假白酒，但酒瘾、心理的压抑、收入的有限和缺乏辨别假酒的能力，使他们完全丧失拒绝消费假酒的自制力。

二、两个村寨的社会文化“病理学”

每一种社会文化对“病痛”都有自身的认知和“病理”解释，或许在此可以借用“病理学”这一生物医学词汇，称之为社会文化“病理学”，以别于基于解剖学的“病理学”。与其他传统社会一样，在两个纳日人村寨里，许多病痛都被视为与鬼神有关。譬如，许多病痛被视为与水井菩萨相关。当村民患病时，其家人往往就会将其归因于水“井”（实指泉水口）菩萨染病后将病痛转给村民。于是，他们需请*da ba*[7]首先确认是村寨的哪一口水井引致得病，然后设坛做法，为那口水井的菩萨驱鬼除秽，以达到为患者治病之目的。其仪式过程繁杂，用时漫长，根据病痛轻重和“病理”差异，用时从数小时至数日不等，其中包括祈求各路神灵和家屋祖先亡灵帮助和护佑。除了向神灵、祖先供奉诸如五谷、糌粑、酥油、牛奶、茶水、蜂蜜等，以及用烧高山大叶杜鹃的枝叶产生的浓烟驱秽治病，用点燃松枝、柏香的袅袅轻烟怡悦神灵和祖先，还要不断周而复始吟唱冗长的经“文”。

在纳日人的“病理”认知中，不得好死，即非正常原因而亡者，其亡灵就会变成萦扰生者的孤魂野鬼。幼儿夜里哭闹、发烧、厌食、出现惊厥等，多被认为是在周围游荡的孤魂野鬼所致，需要请 *da ba* 施法驱鬼除秽。仪式过程中，一方面给孤魂野鬼供奉食物加以慰藉，另一方面颂扬神灵和祖先亡灵的强大威力加以威慑。若出现类似中风、半身不遂、失语、不省人事等病痛，村民则多认为由于天上恶煞所致。其治病仪式更加复杂，且更加费时费力费钱，用于祈求神灵、祖先护佑，慰藉恶煞和孤魂野鬼的供品更多。

纳日社会中，还将一些“病痛”归因于中了“蛊毒”（Weng 1993: 143-167；翁乃群 1996）。笔者在调查中发现，村民所谓中“蛊”的症状多为通常被我们描述为胃肠道和肝胆疾患所致的腹胀、厌食、疲乏体虚、眩晕、消瘦等。纳日社会中有些家户被认为是有“蛊”的。在他们的“蛊毒”信仰中，有“蛊”户的主妇被认为具有对人或牲口放“蛊”之魔力。中了“蛊”毒的人就会患病，并难以治愈。纳日村民认为，西药或中药对治疗中蛊的病痛是完全无效的。据说只有个别 *da ba* 或喇嘛掌握治愈中蛊毒的方法，包括草药制品和治病仪式。一位纳日医生曾告诉笔者，据他了解 *da ba* 或喇嘛用于治疗中“蛊”病痛的草药包含促泻的成分。其实，类似的蛊信仰在许多社会文化中仍在不同程度地流行，并继续对人的社会行为，包括健康实践产生着影响。在田野调查过程中，笔者发现在这些社会文化中都有相似的关于对有“蛊”人和麻风病人采取规避态度的俗语。“蛊”的信仰实为自古以来流传至今、排解无处不在的社会张力的一种社会机制。在现代社会里，它则多被视为“迷信”。人们对“有蛊”人的规避和排斥，则通常会被视为“社会歧视”行为。

虽然在民国时期，生物医学，即西医知识开始传播到纳日人聚居区所属的县城周边，但对于居住在村寨的纳日人来说，他们开始接触到生物医学的时间，大概是在上世纪 60 年代后期。也就是说，在此之前纳日人的“病痛”只是依靠民间传统草药、*da ba* 和喇嘛的祛病仪式来解决。时至今日，纳日人聚居区仍属缺医少药地区，不过大多数纳日村民有了“病痛”一般都会寻求现代的中、西医治疗。值得指出的是，村民在寻求现代医疗的同时，从未放弃原来祛除“病痛”的传统仪式方法。在缺乏 *da ba* 的村寨，原来 *da ba* 祛病除秽的仪式则多由喇嘛仪式所替代。

三、“人观”、信仰与村民的就“医”方式

对于人类学来说，“人”（person）的概念是由社会文化界定的。换言之，不同社会文化或族群都有着自己关于“人”的概念。不同“人”的概念构成不同的“人观”（personhood），以及与此密切相关的不同伦理道德、惯习和行为的重要意识基础。如，川滇边境泸沽湖周边纳日人的传统社会里，“家户领域”是人们最主要的社会导向，而“公众领域”是非常有限的。在他们的“人观”中，每一个人首先是其家屋的一名成员。因此，作为社会一分子，最重要的是当好一生中不同的家屋角色：儿子或女儿、兄弟或姐妹、母亲或舅舅（爸爸）、祖母或祖父等等。维护家屋的兴旺，家屋成员的和睦平安，以及家屋先祖的安宁，是“人”一生的最大祈愿和行为首要准则。

在纳日人的观念中，活人的世界和祖先的世界是紧密相连且互依共构的。前者来自于后者，并始终需要得到后者的护佑，而后者则需要前者不断供奉和祭祀。后者所在的空间是在名为 *Si bu a na wa* 的祖源地的不同村寨，而前者所在的空间则是当下家屋和所在的村寨。这些每每都在 *da ba* 主持的家屋祭祀、驱鬼除秽或治病仪式中被反复地用言语和动作加以表述。纳日人不仅要在自己家屋的经济生产活动中扮演好自己的角色，也要在上述仪式中尽好自己的职责。

纳日人死后（幼年夭折者除外）其亡灵都要通过葬礼仪式被送回祖源地，与先人的亡灵团聚，这是死后其亡灵得以被家人供奉和祭祀，而不会成为无家可栖的孤魂野鬼的必要条件。值得指出的是，只有死在家屋的成员，其亡灵才得以被送回祖源地。在纳日人的信仰中，人死在家屋之外，其尸不得抬回家屋，也不能在家屋里举行葬礼和送魂仪式。与许多社会文化相似，纳日人将死分为“好死”和“不好死”两种。对于因衰老病故或通常认为的正常死亡，只要是在家屋里去世的，均被视为“好死”。对于非正常死亡（诸如因事故、自杀或他杀而死）者的尸体不得抬进家屋，通常就地火葬。这些非正常死者的亡灵也不得被送回祖源地，其名字被排除在家屋祭拜供奉的祖先系谱之外。它们因得不到后人的供奉和祭祀，成为了孤魂野鬼（Weng 1993: 64-66）。正因如此，纳日年长体弱者通常都忌讳离村外出，甚至不愿离家。也正因如此，年老者对上医院看病通常都会有严重的思想障碍，尤其是需要住院治疗时更为困难。当然路途的遥远，交通的不便，以及经济条件也都构成村民求医的障碍，但上述人观和信仰仍是阻碍老人和“重病”患者出远门求医非常重要的原因之一。

四、日本和北美社会对更年期和老年妇女身体的不同认知和医疗实践

加拿大著名医学人类学家玛格丽特·洛克（Margaret Lock）卓有见地地指出，包括生物医学在内的所有医学知识都是以特定的历史、社会、文化、政治和经济为背景，而这些背景对医学知识和实践的本体-认识论（onto-epistemologies）具有着重要意义。她在1993年出版的《老年遭遇：更年期在日本和北美的神话》（*Encounters with Aging: Mythologies of Menopause in Japan and North American*）一书中，讨

论了日本和北美的妇女和医生对妇女更年期的不同医学认知和医疗实践。她指出，自19世纪中叶以来，欧美医学界将女性生命周期的更年期纳入医疗管理的声音越来越强烈。一方面，这与欧美人口平均寿命有较大增长有着密切的关系。在19世纪之前，欧美的人均寿命不超过40岁，对于女性人口来说这就意味着能达到或超过"更年期"年龄阶段的很少。另一方面，人的生命周期变化既包括生物生理方面的，也包括社会角色、地位和社会关系等社会方面的变化，但是长期以来欧美生物医学界在关注人生命周期的不同阶段时，只将个人的身体变化作为聚焦点，而忽视了人的生命周期成长过程中社会方面的变化，尤其是人的主体经历和与之密切相关的社会关系的变化，即身体和社会角色变化之间的辩证互构关系。

日语中原本没有"更年期"的概念。日语的"kōnenki"一词是在19世纪晚期随着日本与欧洲国家交流的增多才被创造出来的。该词用于翻译欧洲的climacterium（更年期）概念。上世纪80年代初，日文著作《妇女与草药》的作者Hideo Nishimura认为"kōnenki"应可用于指不分年龄和性别的任何人的所有生命周期的转折期。值得指出的是，Nishimura的上述意见与19世纪中叶之前欧洲人赋予climacterium一词的意义非常相近。那时在全欧洲，climacterium一词都是指不分性别和年龄的人们生命周期中关键转换阶段相关的各种危险（Lock 1993: 352）。

洛克通过对已绝经一年或以上的45岁到55岁之间的日本妇女关于更年期身体症状的问卷数据和访谈材料进行分析，发现有24%的样本显示她们没有对绝经有特殊的身体症状体验。日本妇女的中老年转折期的身体感受与日本男子的中老年转折期的身体感受无大差异。通常被认为最典型的更经期症状"潮热"项，在所有样本中只有10%选择了该项。而在加拿大马尼托巴省（Manitoba）和美国马萨诸塞州

（Massachusetts）所做的同样问卷调查样本中，该项数据分别为31%和35%。也就是说，日本妇女更年期的“潮热”症状身体感受远远低于北美两地更年期妇女。另一个被认为典型的妇女更年期症状的“盗汗”选项，在日本的调查数据中只有4%，而上述北美的两个调查中，该项的样本数据分别为20%和12%。在北美，当中老年妇女出现上述症状时，多去看妇科，而在日本则是去看内科。

在欧美医学界，妇女更年期症状被视为与妇女的绝经紧密相联系。从生物医学角度，绝经是雌激素水平低下所致，而雌激素水平低下又是由卵巢机能的自然衰退引起的。也就是说，妇女更年期症状是妇女身体器官的衰退引起雌激素水平下降所致。从社会性别角度考察，在欧美医学话语中，虽然中老年男性也与中老年女性一样日渐丧失生育能力（据一研究，六十岁男性有百分七十丧失了生育能力，而八十岁男性丧失生育能力的达到了百分之百），但中老年男性生育功能的下降及其丧失并没有被欧美医学界视为身体的“异常”，而被认为是“正常”的。

早在20世纪30年代初，人工合成雌激素就已被生产出来，但直至20世纪60年代中期之前，人工合成雌激素并没有被广泛用于减缓妇女更年期症状。到了70年代，合成雌激素成为了美国最常用的五种处方药之一，作为治疗/减少绝经期妇女“潮热”和血管舒缩症状的主要药物被广泛应用。到了70年代中期以后，北美著名医学杂志上先后刊登了几篇关于雌激素疗法可能增加罹患子宫内膜癌风险的文章，一度导致使用雌激素减缓更年期症状的治疗急剧减少。此后，这种疗法的利弊在医学界被反复讨论，许多研究者和医生认为更年期妇女如不补充雌激素，便难以在余生拥有相对健康的身体。因此，20世纪80年代在北美，补充雌激素又越来越多地被医生用作减缓中老年妇女更年期症状的疗法，以致雌激素又上升为美国最常用的处方药之一。只是通常医生在采用上述疗法时，还会让患者服用孕酮（黄体酮）以抵消雌激素的毒性副作用。与此同时，医生也被要求在给患者使用此疗法时，要充分评估其可能的危害性，以及为患者选择恰当的激素和剂量，并随时了解患者的情况（Lock 1993: 336-337）。在欧美对身体“异常”的中老年妇女的医疗管理中，医生、医药企业和政府是三个最重要的社会能动主体。

当洛克向一位大阪妇科医生询问他是否认为所有妇女都在更年期时经历了身体上的不适或困扰时，得到的回答是：“不是，我不认为是这样的。”在他看来，妇女更年期症状是一种“奢侈病”。他确信在战前没有中年妇女会为此问题去看医生。一位乡村医生断然地说，农村妇女在更年期无暇感受不适和忧虑，对她们来说即使有一点点感受，也是微不足道的，是很容易轻松度过的。在日本，许多人认为，有很多空闲时间而无所事事的、或者说没有明显地为社会作贡献的妇女，往往缺乏她们母亲以及她们上代妇女的意志力和忍耐力，正是这些妇女会感受到更年期的不适和痛苦（Lock 1993: 348）。

从 20 世纪 70 年代初开始，人口老龄化问题也成为了日本政策制定者关注的问题。其中，重要的担心无非是人口的老龄化将可能增加社会负担而引起国家经济的衰退。更何况日本人口的老龄化进程远比欧美国家更加快速，即欧美社会该过程历时约 100 年，而日本只经历了 25 年左右。20 世纪 80 年代末，一些官员就预测按照当时的趋势，到 2025 年，65 岁以上的人口将占日本总人口的 24%。在这些老年人口中，75 岁以上人口将达到 53% 以上。与此相随的是，到 2025 年将有约 225 万日本老人患有老年痴呆症，其中老年妇女约占 67. 5%。届时，老年人口中约有 200 多万人卧床不起，其中女性占 62%。不同于欧美社会将老龄化问题主要聚焦于更年期妇女的数量，日本则更多地聚焦于卧床者和老年痴呆患者。诚然，由于日本妇女人均寿命比其他国家更长，卧床不起和患有老年痴呆症的数量远比老年男性多，所以她们在老龄化问题中更显突出，但卧床和患有老年痴呆症的男性也同样在上述预测中被提及（Lock 1993: 345）。

虽然日本公众在老年人口的照料问题上，希望政府能有更多的承担，但政策制定者近来仍通过对相关政策的修改，试图改变公众的上述想法。1990年日本一报刊发表题为“护理老人的工作落在妇女的肩上”的文章，指出根据先前对在自家护理老年亲人的近五百人开展的调查，发现家中提供护理工作的主要亲人，81%是平均年龄为56岁的女性。她们中60%的人已照料自己的亲人长达三年以上，其中16%已照料了十年以上。20世纪80年代末一项由日本妇女开展的研究表明，按照当时日本施行的低水平社会保障政策，此后护理老人的负担必将落入到中年女性的肩上（Lock 1993: 346）。也正因为这样，日本更年期妇女无暇感受许多欧美更年期妇女诸如“潮热”、“盗汗”的症状或身体反应，或说她们身体变化的感受与处于同一生命周期阶段的中老年男性无大差异。即使年事已高，只要没有患老年痴呆或其他重病，通常日本妇女在家中继续管理家中经济，操持包括照料和教育学龄前儿童在内的家庭事务。在日本社会性别观念中，妇女的终身社会责任就是养育照料家人。

战后日本，妇女多数被排除在全职工作之外。在“白领”或专业岗位工作领域，很少见到已婚妇女。她们怀孕后，通常都会因社会压力而辞去全职工作。对于她们，当孩子长大后想重新寻找全职工作，通常是很困难的。虽然在当下日本“全职主妇/太太”所占的比例还不足30%，许多妇女日常仍保持忙碌的工作，但所从事的工作绝大多数是不享有社会福利和保障的“蓝领”兼职工作。传统社会文化观念并不鼓励她们这样去做。照料子女、丈夫，尤其家中年事已高，行动不便，或丧失自理能力老人的生活，是中年妇女被赋予的“天生”职责。对于她们，身处“家户”是正常，身处“社会”是异常。在当下政府的公开计划中，也是通过积极鼓励她们负起在大家庭中照料老人的无报酬“天职”，以保持她们不进入劳动力市场。

正是基于上述研究，洛克在该书中提出了具有开拓和启发性意义的“local biologies”概念。北美和日本的妇女对绝经引起的身体变化的主观体验是很不同的，正是这种不同深深地影响着相关话语的建构。在北美社会，老年妇女被视为因绝经而引起的生物性异常，并将其作为医疗干预的对象，而在日本社会，并不对中年妇女的身体变化给予特殊的医疗关注，却对她们有可能因生理变化而丧失在传统三代同堂家庭中照顾老人的能力感到极大的忧虑。北美和日本两地医学界对妇女更年期持有的不同医学知识以及医疗实践是基于不同的“local biologies”。欧美对中老年妇女的关注聚焦在卵巢的退化和雌激素水平的下降而引起的妇女生物机能“异常”（相对于年轻妇女、育龄妇女，以及中老年男性），进而认为需要进行医疗干预；日本则未将妇女更年期的身体和生理变化视为“异常”，而是将其视为与日本老年男人的身体和生理变化一样的正常和自然生理进程，并且不需要特殊的医疗干预。也正因如此，欧美中年妇女出现更年期症状时，多去看妇产科和外科，日本中年妇女更年期期间出现不适多去看内科或中 / 日传统草医。当涉及中老年妇女的健康，日本医生通常都首先建议要注意健康饮食和增加体育锻炼。日本药物企业也因此并没有像欧美同行那样去大量研发生产诸如雌激素类等治疗更年期的药物。当然，北美和日本的社会性别差异，也深深影响着两地妇女对绝经引起身体变化的主观体验。

五、结语

健康问题从来就不仅仅是身体、疾病和治疗问题，更是包括政治、经济、社会文化和人们对其生存的自然和社会环境的自身感受问题。

因此，健康问题不仅是属于生物医学、中医学（在中国）研究的领域，更是人文和社会科学研究的重要领域。

中国是一个人口众多，疆域辽阔的多民族国家。文化上的多元，地理、气候和自然生态环境的多样，城乡间和地区间社会和经济发展的不平衡，医疗卫生资源分配的严重不平等，构成了“健康中国”建设的重要背景。换言之，中国的健康问题是不具有同质性的。因此，“健康中国”建设既需要开展宏观的社会文化研究，更需要开展许许多多具体的、在地的、微观的和“精准”的社会文化研究。“健康中国”计划的有效性，在很大程度上需要依赖于多学科深度参与的、在地的、微观的和“精准”的社会文化研究。作为关涉民生的社会系统工程，“健康中国”计划在实施过程中，不仅需要政府多部门的参与和合作，也需要广大民众，包括群体和个体的积极参与。

注释

[1] 参见 Callahan D. 1973. “The WHO definition of ‘health’”. Taylor S, Marandi A. 2008. “How should health be defined?” *British Medical Journal*. 337 (7683): 1363-1364.

[2] 参见 Bellieni CV, Buonocore G. 2009. “Pleasing desires or pleasing wishes? A new approach to pain definition”. Ethics Med, 25(1); “Sport, Disability and an Original Definition of Health”, <https: //zenit. org/articles/sport-disability-and-an-original-definition-of-health/ > (February 27, 2013).

[3] 同上。

[4] 参见 Ackerknecht, 1953: p.46; Trostle, 1986: 45-46; 翁乃群 2003：p.84。

[5] 参见 World Health Organization (WHO). 2015. Global status report on alcohol and health 2014, <http: //www. who. int/substance_abuse/publications/global_alcohol_report/msb_gsr_2014_2. pdf?ua=1> p. 178.

[6] 该乡直至 2006 年晚春，只有在乡所在村及其相邻的几个村寨有可以行驶机动车的土路，其余村寨间的交通只有可以让人和骡马行走的非常狭窄的羊肠小道。

[7] *da ba* 是纳日人宗教祭师的称谓，同时被作为纳日人宗教信仰的名称。

参考文献

弗里德里希·恩格斯：
《英国工人阶级状况》，《马克思恩格斯全集》第 2 卷，北京：人民出版社，1957。

翁乃群：
“蛊、性和社会性别——关于我国西南纳日人中蛊信仰的一个调查”，《中国社会科学季刊》，1996 年总第 16 期：第 42–54 页。
“艾滋病传播的社会文化动力”，《社会学研究》2003（5），第 84–94 页。

Ackerknecht, E. H.
1953. *Rudolf Virchow: Doctor, Stateman, Anthropologist.* Madison: University of Wisconsin Press.

Lock, Margaret
1993. “The Politics of Mid-Life and Menopause: Ideologies for the Second Sex in North American and Japan” , in S. Lindenbaum & Margarett Lock (eds.), *Knowledge, Power & Practice*. p.330–363. University of California Press.

Trostle, J.
1986. “Early Work in Anthropology and Epidemiology” , in C. R. Janes, et al. (eds.) *Anthropology and Epidemiology.* p.45–46. Dordrecht, Boston: D. Reidel Publishing Company.

Weng, Naiqun
1993. *The Mother House: The Symbolism and Practice of Gender the Na ze in Southwest China.* unpublished PhD dissertation. University of Rochester, New York.

World Health Organization (WHO)
2015. Global status report on alcohol and health 2014: <http: //www. who.int/substance_abuse/publications/global_alcohol_report/msb_gsr_2014_2.pdf?ua=1>

Inventing Chinese Medical *Paidu*: Market, Embodiment, and Self-Care in Contemporary China

Yanhua Zhang*

Introduction

In the mid 1990s when I was in Beijing doing my first ethnographic fieldwork on the practice of traditional Chinese medicine (TCM) in China, the concept of *paidu* 排毒 (literally expelling/purging toxins) [1] was practically unknown. Although *du* (toxin/poison) is a commonly used concept in Chinese medicine, the combination of *pai* and *du* did not occur during my clinical observations and my interviews with patients and doctors. I do not recall any Chinese medicinal herbs or manufactured herbal products being prescribed then for the function of "*paidu*." A decade later when I began my research on adverse effects of Chinese medicine in Beijing and Dalian, *paidu* had become a common word

* Yanhua Zhang, Department of Languages, Clemson University.

in everyday language. Information on *paidu*-related health products and services were everywhere, from billboards advertising herb-based *paidu* capsules for losing weight and enhancing beauty, TV commercials promoting *paidu* suppositories for women's reproductive health, signs on the windows of beauty salons and health spas offering various *paidu* services, to restaurant menus recommending specific dishes for *paidu* effects.

Within a short time, *paidu* has become a popular healthcare concept and a new modality of preventive health practice, especially among urbanites in China. A large number of the urban population participate in *paidu* practices to varying degrees including consuming health foods, taking medicinal herbs, and engaging in more rigorous therapeutic regimens in order to expel toxins from the body. Initially the concept has an explicit association with *yangyan* 养颜 (enhancing skin beauty); it is therefore particularly popular among female white-collar professionals. However, a former colleague, a middle-aged male college professor, convinced me that *paidu* is also "in" *(re)* among middle-aged men who feel the need to lose weight, keep fit, and retain youthful energy. He himself sought out several *paidu* treatments at a health clinic and claimed to feel healthier and more energetic afterwards. Over the past decade, *paidu* therapeutics have expanded from a few herb-based products to numerous varieties, such as meridian *paidu*, scraping *paidu*, fumigation *paidu*, Chinese medical holistic *paidu* (*zhongyi zhengti paidu*), and lately *paidu* for nurturing life (*paidu yangsheng*).

The media and the popular discourse tend to place *paidu* within the paradigm of traditional Chinese medicine and claim that *paidu* concepts and methods are based on traditional Chinese medical knowledge, which can be traced back to antiquity. Recently, the concept of *paidu* and its ubiquitous practices have become topics of concern in the professional community of traditional Chinese medicine. The concept of *du* and its clinical significance have been popular subjects of theoretical and empirical investigations by scholars and practitioners of Chinese medicine, and the language of *paidu* has been assimilated into the mainstream clinical reasoning and treatment in TCM clinics. In August 2009, *the China Association of Chinese Medicine* (Zhonghua Zhongyiyao Xuehui) and *the Acupuncture Institute of the China Academy of Chinese Medical Sciences* (Zhongguo Zhongyi Kexueyuan Zhenjiu Suo) jointly sponsored a national conference in Beijing on TCM therapeutic methods of *paidu*. The organizers of the conference claimed that *paidu*, as a preventive healthcare strategy, has been widely used in clinical settings as well as in non-clinical health practices at home and in the market, and that *paidu* was in the process of developing into "a new Chinese medical theory (学说)." [2] The conference called for further theoretical and clinical research on *paidu* theories in order to better guide the public in understanding and utilizing *paidu* products and services. The participation of professional TCM practitioners in re-making *paidu* is significant, indicating that a popular healthcare fad, driven initially by market interests and consumer demands, is now not only being assimilated into the domain of institutional Chinese medicine, but also researched and legitimatized as an emerging TCM modality of treatment and theory. In this process, traditional Chinese medicine itself is undergoing "reassembling" in response to social, economic, and material changes of everyday life in contemporary China.

As is true everywhere in the globalized world, science and technology are conveniently enlisted in the service of "the tradition." [3] The *paidu* therapeutics is frequently presented as the product that combines traditional Chinese medical knowledge and materials with the advanced biotechnology from developed countries for the care of the modern cosmopolitan bodies indulging in excessive consumption of all kinds and exposed to all sorts of environmental pollutants and health hazards. [4] This fusion of the tradition and advanced technology creates space for negotiation between sets of conceptual and practical contradictories such as accessibility and risk, healing power of the tradition and the epistemological authority of science, the natural holistic approach and advanced biotechnology, etc.

Then, if bodies have histories or rather "remember" (Fassin 2007) and behave "in new ways at particular historical moments" (Csordas 1994: 1), what does the popularity of *paidu* tell us about the particular historical moment that makes *paidu* a necessary or desirable part of embodiment in everyday life, especially when taking into consideration the magnitude of *paidu* products and services consumed in the urban area? Furthermore, how did the concept of *paidu*, a contemporary invention and likely an introduction from outside China, achieve such a status as both traditional and Chinese? A particular focus is on how the diverse personal, cultural and material agents are motivated and assembled to forge the link between *paidu* and the traditional Chinese medicine, thus transforming both the *paidu* practice and traditional

Chinese medicine. Finally, what motivates ordinary urbanites to engage diligently in *paidu* practices, though they could be economically costly, time consuming, and sometimes physically unpalatable? In other words, I am interested in what consumers of *paidu* gain from their experience of practicing *paidu* and what subjectivity such a health practice and consumption express and produce. By focusing on the invention of *paidu* therapeutics as a topic of ethnographic inquiry, it allows me to explore the complex interplays that link the human bodily with the historical and material as well as the moral and personal in making the worlds in which contemporary Chinese urbanites live.

The information upon which this paper is based comes from multiple sources. First, it draws from my own observations and participatory engagement in everyday life in China, mostly in the cities of Dalian and Beijing during the summers from 2009 to 2015 while I was coordinating summer study abroad and internship programs. As a true participant, I was able to observe and interact with doctors and patients in a major hospital of traditional Chinese Medicine and in a community health center specializing in traditional Chinese medical therapies. Occasionally, my own body became an ethnographic tool receiving *paidu* massages in Chinese medical clinics. Second, health-related TV programs such as "*Zhonghua Yiyao* 中华医药 (Chinese medicine)," "*Jiankang Fangtan* 健康访谈 (Health Interviews)," and "*Jiankang Zhi Lu* 健康之路 (the Way to Health)," and numerous healthcare-related websites were rich sources of information. I paid particular attention to the social media, which have emerged as popular venues for circulating health information and knowledge and for sharing experience and ideas.

Some blog sites have grown into relatively stable cyber communities where people address each other with kinship terms, sharing personal stories, and offering each other moral and affective supports. Finally, the published research and case studies on *paidu* are valuable sources of data for understanding the expert knowledge and professional voice in shaping *paidu* practices and their manifold meanings.

In the following, I will first historicize *paidu* and situate its emergence and development as a popular urban preventive healthcare movement in the context of the post-socialist market-oriented society of contemporary China. I will then trace the changing meaning of *du* in traditional Chinese medicine and show that the assimilation of *paidu* into the discourse and institutional practice of Chinese medicine not only re-defines the *paidu* concept, but, to certain extent, reconstructs the "objects" to which Chinese medicine attends and thus also transforms the way Chinese medicine is practiced in the clinic and utilized in general. This inquiry into the process of re-making *paidu* informs the understanding of Chinese medicine as a living tradition that has been constantly reassembled, regenerated, and appropriated in relation to the changing sociocultural and material conditions. Finally, using a blogger's narrative of her *paidu* experience and reflections as an example, I will discuss how *paidu* therapeutics is appropriated as a "technology of self care" for managing and achieving bodily and social health in the face of changing social and moral contexts in contemporary urban China. This later development of *paidu* has been integrated into the popular *yangsheng* 养生 (life nurturance) movement and become known as *paidu yangsheng* or *yangsheng paidu*.

Marketing *Paidu* for the Excesses of Modernity

Pai (expel/purge) and *du* (toxin/poison) are two commonly used terms in everyday Chinese language, and *du*, in particular, is an important Chinese medical concept concerning medicinal effects of herbs and illness etiology. The combination of the two, however, was a recent creation. *Paidu* wasn't a known concept to the general public until after 1996 when the Chinese herb-based product *"Paidu Yangyan Jiaonang"* 排毒养颜胶囊 (capsules for expelling toxins and enhancing skin beauty, henceforth, *paidu* capsules), manufactured by a pharmaceutical company based in Yunnan province, was first introduced to the health product market. [5] Larry Lang (2008), a professor of finance from the Chinese University of Hong Kong, is probably the first person to point out that *paidu* is "an extraordinary creation" of a pharmaceutical company. As an economist, Lang attributes the commercial success of the *paidu* capsules to the extraordinary creative marketing strategies of the company.

According to Lang (2008), the first *paidu* herbal product became an instant business success for the Yunnan-based pharmaceutical company. [6] The *paidu* idea then quickly captured the public's imagination and became everyday vocabulary among urban residents. As cited in his book, the consumers' reports conducted in the cities of Shanghai, Hangzhou, and Wuxi in 2002 showed that the majority of consumers became more concerned about toxins in their bodies, and that 48. 5% of consumers using *paidu* products in general gave a positive evaluation. [7] The success of the first *paidu* capsule inspired many in the health product industry to follow suit to produce the second and third generations of *paidu* products, such as *luhui paidu jiaonang* (芦荟排毒胶囊)(aloe *paidu* capsules) and various *paidu* teas. The concept of *paidu* continues to gain popularity, and vegetables, fruit, and grains are often recommended for their *paidu* functions in restaurants and in countless food-related commercials. There is even a joke to the effect that when the Chinese greet each other, instead of asking "Have you eaten yet?" they now ask "Have you *paidu* yet?" [8]

Can this massive success of *paidu* products and the popularity of the concept among healthcare consumers be primarily attributable to a well-designed marketing strategy that dresses a new concept with the familiar language of traditional Chinese medicine, as Lang (2008) claims?Traditional Chinese medicine recognizes both deficiency and excess as illness factors, and replenishing (补) and clearing/draining (清 / 泄) as two corresponding corrective strategies. In general, the preventive healthcare market has been dominated by the replenishing strategy with sorts of herbal tonics for strengthening and nourishing various bodily-affective systems. Such practices conform to the therapeutic wisdom of traditional Chinese medicine, which teaches that the vital substances of the human body, such as *qi* (气), blood (血), and essence (精), tend to be gradually consumed and weakened when the functional systems of the body are not sufficiently nourished or regenerated due to aging, chronic illnesses, malnutrition, and physical or mental exhaustion. [9] This explains why Chinese herbal tonics and medicinal foods make good gifts for aging relatives and recuperating patients.

However, the sudden boom, in the 1980s and the early 1990s, in consumption of replenishing herbal products and medicinal foods as an universal practice for enhancing health and preventing diseases was recognized as extraordinary, very likely a post-socialist phenomenon. [10] Judith Farquhar (2002: 129) notices "a marked emphasis on medical technique of *bu*" for "bolstering the depleted or hypo-functional body" in China's reform era. According to Farquhar, the fervid desire in seeking pleasure can be understood "as a way of making up for the poverty and misery many associated with their collective history." In other words, bodies have memories that are constantly interpreted and appropriated for understanding and making the present. For many Chinese, preventive healthcare fixing on consuming herbs and medicinal meals to replenish deficiencies may well be rooted in the collective bodies that memorized food deficiencies and restricted consumption in China's pre-reform era. [11]

If the obsession with replenishing consumption was motivated by an imagination and the memory of pathological deficiencies in need of nourishment, then the recent booming business of *paidu* is likely motivated and created for a different bodily awareness and experience in the everyday life replete with material goods. *Paidu* casts the condition that compromises the health as "excess" rather than "deficiency", which requires a therapeutic approach of "expelling or purging" instead of "replenishing". For the preventive health product addressing the pathological problem of excess to replace the traditional tonics in securing a large market share and consumer support, there must be complexities beyond an effective marketing strategy, which itself reflects the underlying changes at multiple levels—the social and economic, the cultural and ideational, and the bodily and personal—in post-reform

China.

From the mid 1990s into the new century, China's economic reform was stepping up and culminated with China joining the WTO (World Trade Organization) in 2001. The large-scaled privatization of state-owned enterprises, the marketization of healthcare and education, and the increased presence of foreign direct investment have drastically transformed cities in China and everyday life in China's urban space. Larger cities witnessed the emergence of a new social stratum of white-collar business professionals employed in multi-national corporations owned by foreign investors, and with it came "a lifestyle and worldview that is emblematic of global capitalism as it is found in contemporary China" (Duthie 2005: 2). Embracing and aspiring for such a "modern lifestyle" are also professionals working in state-owned enterprises and in the private sectors as well as those educated elites employed in academia and government institutes. The rising middle-class urbanites are now the major consumers and clientele of preventive healthcare products and services. Deficiencies are still found in urban spaces among the unemployed or underemployed, most of whom are laid-off workers from the previously collectively owned or state-owned enterprises and migrant workers from the rural areas. However, with limited consumption power, they are less the targeted consumer groups of the health product industry, and their deficiencies are beyond the power of Chinese medicine or any consumer medicine that focuses mainly on the individual bodily.

Clearly, emergence of *paidu* coincided in many ways with the rise of the urban middle-class and a type of lifestyle it embodied, and was specifically promoted as preventive health care strategies for those living a busy, sedentary, and economically comfortable urban life. The targeted consumers of *paidu* products and services initially were white-collar workers or professionals who spent long hours sitting in their offices and in front of computers and business owners or officials, who tended to banquet excessively to close deals. These potential consumers were advised to watch out for the symptoms of accumulation of "*du*" , typically manifested as skin problems, constipation, overweight, and other related sub-health problems. The *Panlong Paidu Capsule*, for instance, indicates that the product treats the symptoms of constipation, acne, and facial pigmentation by expelling toxins from the systems of the body. [12] As *paidu* gained popularity, more advanced and sophisticated *paidu* services became affiliated with established traditional Chinese medical institutes, such as the *Paidu* Center at Guangzhou *Zijing Hospital*, which offers to completely clear away *du* in a client's body so that males will enjoy a strong healthy body and be confident, and females will enjoy health, beauty, and youth. [13]

Then, what is *du* and how did it become this inevitable derivative of modernity that poses a threat to well fed and nourished cosmopolitan bodies?

The Panlong Yunhai at its company's website offers the following introduction:

> Urban residents are now living in a heavily polluted environment. They are increasingly exposed to toxic substances produced by environmental populations due to rapidly increased numbers of

> cars and factories as well as excessive use of chemical fertilizers, pesticides, and food additives. These externally generated "*du*" are harmful to human health. Furthermore, with the continuous improvement of living standards, people tend to consume excessively including large quantities of nutritiously rich and greasy foods in their diet, which can result in disordered metabolic functions and accumulations of metabolic wastes in the body to form endogenous "*du*" . ... These toxins generated externally and internally, if retained inside the body, can lead to malfunctions of various organs and tissues resulting in an imbalanced flow of *qi* and blood, which triggers a variety of diseases manifested as constipation, acne, facial pigmentation and other symptoms. [14]

Obviously, *du*, as explained in the context of *paidu*, is attributed to various forms of excesses. The externally generated *du* coming from all sorts of pollutants—industrial and automobile waste and emissions, agricultural pesticides and fertilizers, chemicals in food and cosmetic products— are "excesses or side effects of modernity." [15] Similarly, an article from the website of the Chinese Medical Center for Natural and Holistic *Paidu* affiliated with Guangzhou Zijing Hospital asserts: "Living in such a polluted environment, ... our body cannot bear the continuous influx of *du* and loses its physiological functions, and to restore good health we must restore the body's *paidu* functions by first expelling *du* from our bodily systems." The author then asks, "When did you give your body a maintenance and clean your body's filters (as you do for your car)?" [16]

The endogenous *du* in the form of pathological repletion in the Chinese medical sense resulting from excessive behaviors such as unruly eating or indulgence in caffeine, alcohol, and nicotine is another form of excesses attributable to the modern lifestyle. In a sense, everyone living a modern life is exposed to "*du*" or is carrying *du* in his or her body. The commercials for the *Panlong Paidu Capsule* claim that a normal adult carries in his or her body about 5. 6 kilograms of *du*, and according to "the Basic Knowledge of *Du*" , the weight of *du* in an average adult's body ranges between 3 and 25 kilograms. [17] This explanation of *du* and *paidu* plays into the concerns of ordinary Chinese who in their everyday lives are increasingly exposed to environmental deterioration and food contamination, especially after several publicized high profile scandals concerning food safety such as the incident of "poisonous milk powders" in 2008.

Regarding the pervasiveness of *du*, modern bodies are permeable and *du* does not recognize boundaries between excess and deficiency. But, *paidu* as an embodiment practice does. Commercial and preventive medical *paidu* could be costly. Where to go for *paidu* or what *paidu* products and services one utilizes is very much a matter of social status (身份) as I was informed in the summer of 2009 by a friend, a typical *bai-gu-jing* (白领 / 骨干 / 精英 white-collar/important/elite) working at the Beijing Financial Street. The idea of *paidu*, though widely spread and embraced by many urbanites now, is nevertheless linked, at least initially, to "a kind of life mode and attitude", [18] that values individual choice and personal responsibility leading "to enhanced health, and finally to a marketable self" . [19]

In this sense, the emergence of *paidu* in the healthcare market was not simply an introduction of a new health product, a new preventive healthcare regime, or a new concept of consuming Chinese medicine. From "replenishing deficiencies" to "draining excesses" , *paidu* has emerged rather as a new technique of self-making and a model to embody health and well-being in the wake of neoliberal modernity. As claimed by the advocates, *paidu* has become a lifestyle, both healthy and fashionable, emblematic of excess and status.

Mainstreaming *Paidu* into Professional Chinese Medicine

Although *paidu* emerged in response to the excesses of modernity and to the need of the rising urban middle-class for a new modality of preventive care, from the very beginning, the *paidu* rhetoric associates the therapeutic concept and practice with traditional Chinese medicine. Then, what role did traditional Chinese medicine play in its invention, production, and legitimization? How did the process of "networking" *paidu* into TCM theories and methods transform the *paidu* concept and its use, and at the same time, "reinvent" Chinese medicine in ordinary clinical practice so that it stays on track with the changing socioeconomic, bioenvironmental, and technological worlds, and with changing health needs and consumer demands?

As the first generation of *paidu therapeutics* appeared in the market, the health product industries made efforts to align their products and services with traditional Chinese medicine and actively promote them as "Chinese" , "traditional" , and "natural" . The motivations were partly strategic. [20] Manufactured *paidu* products as marketed are likely perceived as comparable to herbal tonics and used by consumers with little concern for side effects. Even when the ingredient, Chinese rhubarb (大黄), which is commonly used in *paidu* products, can cause stomach discomfort and diarrhea, such an effect is likely to be perceived and interpreted positively as an indication of potency of the product or a sign that the remedy is working. The deliberate marketing of a *paidu* product as traditional and herb-based not only gives the product a therapeutic appeal as a natural solution instilled with ancient cultural wisdom to health problems in the modern time but also links the newly coined concept of *paidu* in material to the ancient tradition of Chinese medicine.

In addition, the famous slogan used to promote *paidu* capsules "*da tong paidu guandao* (opening up the *paidu* channels)" conceptually resonates traditional Chinese medicine. From the perspective of Chinese medicine, the bodily is conceived in terms of process and transformation. Health depends on a dynamic balance of bodily functions or the orderly flow of *qi* and other life forces. To facilitate the orderly and unobstructed flow of *qi* is the way to maintain health. This "unobstructed flow" is characterized as *tong* 通 (opening up, flowing, connecting) (see Zhang 2007). In a way, highlighting "*tong*" as the therapeutic mechanism behind *paidu* draws it conceptually close to traditional Chinese medicine. When the practitioners and scholars of Chinese medicine later adopt the concept of *paidu*, they readily embrace the expression of "opening up the *paidu* channels". [21]

However, as *paidu* aligns itself with traditional Chinese medicine for its market appeal, it has also invited a closer scrutiny from the community of professional Chinese medicine and the competition for the market share. Panlong Yunhai Pharmaceutical Company claims the formula of their *paidu* products embodies the Chinese medicine's practice of "differentiation of patterns and determining treatment (辨证论治)" —the heart of the contemporary practice of Chinese medicine, and that the health product integrates the "four different treatment methods of traditional Chinese medicine—*tong* (opening up), *jie* (dissolving), *tiao* (harmonizing), and *bu* (replenishing)." [22] Ironically, these characteristics of traditional Chinese medicine are exactly what a manufactured herbal product lacks. The principle of "differentiation of patterns and determining treatment" is embodied in the clinical practice of Chinese medicine that emphasizes particularities in patterns of illness manifestation in a particular patient and in treatment strategies, which is beyond what a uniformly designed and mass-produced health product can accomplish. Furthermore, clinically practiced Chinese medicine does use different treatment methods, but they tend to be strategically interwoven into the temporal unfolding of illness development.

Understandably, the participation of professional Chinese medicine in the making of *paidu* started with the criticism of ready-made *paidu* products. For example, an article posted in January 2005 on the "39 *Jiankang*" , the popular Chinese health related website, cited Professor Du Guiyou from the Chinese Academy of Traditional Chinese Medicine:

> From the perspective of traditional Chinese medicine, it is inappropriate to conflate the strongest replenishing herbs and the strongest draining herbs in a single treatment formula as the *paidu*

capsule does. In the practice of traditional Chinese medicine, this strategy may be used, but only for an extreme case in an urgent situation. But the *paidu* capsule is recommended for consumers for treating non-acute common symptoms, such as "constipation, acne, and facial pigmentation" . People with Chinese medical knowledge know immediately that this is a problem, but ordinary consumers with no such knowledge would take the capsule regularly. This would definitely cause health problems. [23]

Practitioners of traditional Chinese medicine question the widely advertised *paidu* products and reject the herbal regimens sold for draining excesses as an authentic Chinese medical practice for managing *du* infused bodies. Contrary to what many consumers believe, *paidu* is not an original Chinese medical notion with a history of thousands of years. In fact, to combine the verb *pai* (to expel) with the noun *du* (toxin/poison), implies a revision in the meaning of the Chinese medical concept of "*du*" , and coming with it a revision in its approach to "*du*" . "*Du*" in Chinese medicine is not an equivalent to the English word toxin or poison, which by dictionary definition is a disease-causing antigenic substance. The original meaning of "*du*" found in the ancient Chinese dictionary was "thick/thickness" .[24] In *Huangdi Neijing* (*Inner Classic of the Yellow Emperor*), "*du*" and "yao" (medicine/herbs) were not completely separate concepts, and they were often used in combination, such as *du-yao*, to refer to the medicinal plants that induce strong reactions in the body. *Du* had the meaning of "strong" , "violent" , and "intense" in reference to potency of medicinal plants. [25]

Later in *Zhubing Yuanhou Lun* (*Discussion on the Origins of Symptoms of Illness*) published in the Sui Dynasty (581–618 A. D.), the meaning of *du* was expanded to describe *xie* 邪 (pathogenic factors). Since then, *du* and *xie* have been commonly used in combination to refer to disease-causing factors that are particularly harmful and detrimental to the bodily systems and functions. Here *du* has retained its semantic root denoting intensity and degree. When a pathogenic factor is perceived as particularly severe, infectious, or resisting to treatment, it becomes *du* as in the expression "邪盛为毒 *xie sheng wei du*" (Pathogenic factors running rampant become *du*.) (Zhao 2007: 7). Apparently, at least until recently, *du* in Chinese medicine, though polysemic in nature, has been understood more in association with force and power that affect or activate changes in the body, rather than a concrete, tangible substance external to the body. As Yu Zhimin (2007: 45) points out, "to assume that *du* simply refers to some tangible material substance is to miss the basic point of Chinese medical approach to treating illness" . This also explains why "*du*" in Chinese medicine routinely takes up the verb "*jie*" (dissolve) to represent the treatment approach to "*du*" , which has the meaning of "dissolving" and "mitigating" the intensity of the harm through the strategies of easing, neutralizing, or transforming the pathogenic force.

On the other hand, in the Chinese medical context, the verb *"pai"* tends to be associated with concrete substances—wastes to be discharged from the body, such as *pai bian* (excrete) and *pai niao* (urinate), which can be generally grouped as the method of *xie* 泄 (purge or drain). The combination of *pai* with *du* in fact remakes the concept of *du* from something intangible, transformative, and dynamic, into the tangible waste matter that can be cleansed out of the body. Noticeably, many manufactured *paidu* products do contain the herbal elements such as *dahuang* (Chinese Rhubarb) and *fanxieye* 番泻叶 (Folium Sennae) as laxatives to facilitate bowel movement. Such function is well captured by the popular advertisement for a certain *paidu* product "to give your intestines a bath". The following *paidu* promotion is also typical: "A car needs a maintenance every 5000 KM, which includes replacing the air filters, oil filters, fuel filter, and more... When do we give our body a maintenance and wash the body's filters?" [26] Clearly *paidu* as promoted in the commercials not only renders the body lifeless and mechanical, but constructs the "body-object", which the *paidu* product is invented to attend to and care for. Using the Nina Etkin's words on the health food industry, the claims of *paidu* "have more to do with increasing market share than with discovery of health potential of such products" (2006: 213).

For the professional TCM, to claim the ownership of *paidu* and to reinvent it as a legitimate Chinese medical theory and method that meet the health needs in China today, some serious work is needed, both theoretical and clinical. This is the context for the emergence of a large number of TCM researches that make "*du*" and its treatment the focus of the investigation (e.g. Yu, Junsheng 2000; Yu, Minzhi 2007; Zhao, Zhiqiang 2007). In the preface to the book by Yu Zhimin (2007)that focuses exclusively on the evolution of the concept of "*du*" in traditional Chinese medicine, Wang Yongyan, the honorary president of the China Academy of Traditional Chinese Medicine, commends the research as "theoretical innovation", which "is a necessary step (to bring Chinese medicine) from traditional to modern".[27]

For many TCM researchers, the problem of how to work the contemporary meaning of *du* as understood by lay persons and represented by the *paidu* industries into the existing traditional Chinese medical corpus becomes the task of how to reinterpret the Chinese medical concept of *du* and *duxie* to accommodate *paidu*. Often "*du*" is discussed in light of other pathogenic concepts, such as *tan* (phlegm) and *yu* (stasis) that have both tangible and intangible aspects. The tangible "*du*" can include the meaning of toxic substances and effects generated externally by various environmental pollutions and internally by impaired bodily functions. [28] Then, taken and explained as a pathogenic factor of various toxic substances generated within or without human bodies, *du* is literally *duxie* 毒邪— the concept that is dealt extensively in the Chinese medical classics. On the other hand, *du* is also interpreted as "intangible" and retains its semantic root as a relational concept of degree and intensity; therefore *du* could also be any illness factor (邪) including any form of "tangible" toxic substance that is on a rampage, extremely unstable and resistant to treatment, and deadly in its capacity to affect the bodily systems (see Yu 2007). According to Yu, *du* in Chinese medicine can be taken as a theoretical model (*lilun gongju*)to understand the complicated relations between illness factors (病因), illness mechanisms (病机), and material and environmental contingencies. [29] *Du* renarrated as such is multifaceted and inclusive, which accommodates both the biomedical categories of toxics and the layperson's common sense notions of pollutants and accumulated wastes inside the body.

Once *du* is defined as a Chinese medical concept of *duxie*, the long history of Chinese medicine involving tackling "*duxie*" becomes relevant, especially the school of "repulsing pathogenic factors" (攻邪) with various draining and purgative methods, and *paidu* can be legitimized as one of those. (see Zhao 2007: 12–13) It follows that *paidu* as a Chinese medical therapeutic approach has to be based on a careful diagnostic process of *bianzheng lunzhi* (differentiation of patterns and determination of therapies) that takes into consideration of all the particularities of a specific individual and the social and material contexts to decide on a particular treatment or *paidu* strategy. In this sense, authentic *paidu* therapeutic practices for treating illnesses and for preventive "attuning" are beyond beauty salons and health spas, and are more complicated than just taking manufactured capsules. The years following the new century have witnessed the movement of *paidu* into TCM hospitals and clinics.

With *paidu* re-narrated and assimilated theoretically into institutional Chinese medicine, changes in clinical practice are easily observable. *Du* is very often made the focus of the clinical action for deploying therapeutic resources and techniques in treating ordinary medical disorders. Some forms of *paidu* procedure are routinely integrated into conventional treatment strategies to drain the accumulated excesses in the body and to unblock the bodily channels before starting standard therapies. For example, a conventional treatment of female infertility in traditional Chinese medicine's Women's Specialty (妇科) primarily involves the visceral systems of the kidney and the liver since, based on the physiology of Chinese medicine, the kidney stores *jing* (essence/

semen) and the liver stores *xue* (blood), and sufficient *jing* and *xue* are the basis for pregnancy. Doctors tend to regard the majority of infertility cases as the result of kidney depletion or liver constraint and frequently use warming, replenishing, and nourishing strategies to correct the kidney depletion and activate the liver blood in order to help patients achieve pregnancy. [30] However, my observations and subsequent interviews in the Municipal Hospital of Chinese Medicine in Dalian in the summer of 2009 reveal that *"du"* became a central concern in the treatment of infertility and in many cases the treatment plans partly involved *paidu* procedures, which is clearly innovative.

The patient whom I followed closely was an acquaintance in her early 30s. [31] She started to receive treatment for her "woman's problem" (妇科病) in the summer of 2009. When I was interviewing her, she had just finished her first course of treatment and was about to begin her second 10-day treatment course. The first two courses could be characterized as *paidu*, involving different types of internal cleansing and draining, washing, and massage. After the first two courses of treatment, she would then receive her herbal prescription.

She reported after the first course of *paidu:*

> "I finished the first 10 days of treatment, mostly *qingchang* (colon cleansing) therapy. There was no adverse effect on the body. The treatment cleared the body of *du*, and it also had cosmetic effect."
>
> "One course of treatment cost less than 500 *yuan* (approximately 75 US dollars in 2009). I feel good, better than going to a beauty

salon, and I am also going to have *anmo* (medical massage), once a week."

"The symptoms of inflammation were improving day by day. After the first course, my inflammation was almost gone completely."

The patient was familiar with the *paidu* concept and mentioned repeatedly how the treatment was natural and did not have any adverse effects to her body. Her type of treatment that involves *paidu* procedures has become a new norm in the treatment of gynecological problems in this clinic. Once I asked why so many young women had *fu ke* problems, when commenting on the large crowd waiting outside the "*fu ke*" consulting room, she answered "too much nutrition, and toxic chemicals have no way out, and all get stuck in a fallopian tube."

Later, she sent me via instant messenger the herbal prescription she received from the doctor, which contains 14 different herbs. The main ingredients, 当归 (Angelica), 川芎 (Ligusticum), and 白芍 (Paeonia lactiflora), are the common ingredients of the formula for treating gynecological problems through nourishing blood and harmonizing the liver (养血调肝) as well as activating blood and moving *qi* (活血行气). This case clearly shows that the idea and mechanism of *paidu* has been incorporated into the clinical reasoning in the practice of Chinese medicine in an institutional setting. Here *paidu* is used as an addition to the temporal form of treatment to ready the patient rather than to completely change the therapeutic logic of treating infertility. Recent studies on *duxie* theory and its application in clinical work demonstrate that many common disorders in a clinic of traditional Chinese medicine are now approached from the perspective of *duxie* and their treatment involves some procedures of *paidu*. [32]

Another observable change following the incorporation of *paidu* into the institutional practice of Chinese medicine is the branching-out of Chinese medicine into "new" territory—a preventive medicine epitomized by the establishment of the TCM *Paidu* Center such as the one established in the Guangzhou Zijing Hospital of Chinese medicine in August of 2009. [33] Although, Chinese medicine is always talked about in terms of illness prevention and believed to be particularly resourceful for treating the "not yet ill (未病)" or "the sub-health (亚健康)" in the contemporary sense, in reality modern TCM hospitals do not normally have a division exclusively devoted to preventive medicine. Preventive care is either left to self-health practices outside the TCM institutions or integrated into the conventional clinical practice.

The preventive healthcare division, such as the TCM *Paidu* Center at Guangzhou Zijing Hospital and the Sub-Health Department in Beijing Dongfang Hospital, is a recent innovation indicating a change from a treatment-based care to, more or less, a market-based one. Although the TCM Holistic *Paidu* Center at Guangzhou Zijing Hospital is labeled as TCM and claims to provide individually designed holistic *paidu* treatment prescribed according to the diagnostic principles of Chinese medicine, it is truly a hybrid and cosmopolitan institution where one can find ideas, products, and techniques imported from different parts of the world, including naturopathic medicine from the West, ideas from Indian Ayurveda medicine, and products and techniques from South Korea and Japan. According to the contract between the Center and the consumer, the cost for a complete *paidu*, which could last up to 12 hours, costs as much as 3, 000 Chinese yuan (approximately $430 US dollars in 2009). Here at the TCM Center, the holistic *paidu* belongs to a different kind of "Chinese medicine" where bodies of patients are made into bodies of consumers and clientele of desirable commodities—optimal experience of health by choice.

Paidu as a Form of *Yangsheng* (Life Nurturance)

As *paidu* is formulated as traditional and Chinese and extending from the market place to the TCM clinics, it has also been personalized and practiced at homes as self-health and a new form of life nurturance (*yangsheng*), [34] which is popular among the well-informed urbanites who are familiar with the rich narratives and practices of *yangsheng* and also instilled with the post-socialist neoliberal ideology that can be translated as "my health, my say" (我的健康我做主)—the slogan of a popular CCTV 4 health interview show. Increasingly, *paidu* is promoted and interpreted in terms of nurturing life, and the combination of *paidu* and *yangsheng* has become a constant topic featured in the self-health media and popular TV health programs. For instance, "*Yangsheng Tang* (The Hall of Life Nurturance)," a popular BTV health show, ran a series on *paidu yangsheng* to advocate *paidu* as a new strategy of life nurturance. [35] Now *paidu* has been effectively assimilated into the magnitude of contemporary *yangsheng* repertoire. [36] In fact, *paidu yangsheng* or *yangsheng paidu* sounds so familiar to the ears of ordinary Chinese that the combination of the two seems to be only natural.

While it is true that both *paidu* and the contemporary *yangsheng* movement claim their roots in traditional Chinese medicine, yet the typical discourse of *yangsheng* as nurturing life and *paidu* as expelling pathogenic excess suggests two different strategies of care. If the contemporary *yangsheng* is to mobilize positive potentials of life through actively participating "in the natural power and virtue of giving life to life" (Farquhar & Zhang 2012: 14), *paidu* centers on targeting the adverse bodily conditions by flushing out toxic accumulation that is alien to the body through intimate bodily manipulation or ingestion of medicinal products. The combination of the two is rather creative than natural and can be seen as "tactics of the everyday" in response to the diverse needs and experience of urban populations in their search for a life that is physically and mentally healthy and also culturally and morally good and meaningful in the much changed sociopolitical, material, and moral contexts of today's China. Then, who are the people engaging *paidu* as life nurturance, and what is the significance of associating *paidu* with *yangsheng*? Do they approach *paidu* differently from those seeking conventional *paidu* for "giving the body a general cleaning" [37] or clinical *paidu* as a therapeutic procedure? Also, is the *yangsheng* movement reinventing or expanding itself by taking up *paidu* as a way of nurturing life? The narratives by those engaging *paidu* for nurturing life may offer useful insights.

Xueluo Huangcun 雪落荒村 is an online ID of a middle-aged female professional, who started to practice *baguan paidu* 拔罐排毒 (cupping *paidu*) at home since August 2009 and later tried other forms of *paidu*. [38] She uses cupping *paidu* for self-health and also in treating her family members, while blogging about the revelations and experience gained in the process. She has a group of friends who interact mostly online, and is often referred to as "Sister Snow" in their online communication. They share information on everyday topics, such as health, child rearing, schools, books, etc. Sister Snow clearly understands her cupping *paidu* practice as a form of life nurturance. In responding to a comment on her *paidu* post, she wrote on 16 August 2009:

> With young people in their 20s and 30s, do not talk about *yangsheng*... Once reaching the middle age, like tree leaves feeling the autumn, we start to feel that our body is weakening, and the tolls of years of overwork on the body start to manifest as various illnesses; then we realize that health is most important. My purpose for learning *yangsheng* is not to seek to extend the absolute length of life, but to improve the quality of living for the second half of my life, hoping every day I live will be a healthy day."

As shown earlier, when the first *paidu* capsule was invented in the mid 1990s, the concept and the product were associated with the notion of *yangyan* (养颜), which was marketed primarily to the white-color workers and professionals, young or in their prime, for the purpose to cleanse the body and maintain healthy facial complexion and youthful energy. From *yangyan* to *yangsheng, paidu* is now quite popular among the generation who are past their prime or, in Sister Snow's words, "feeling the autumn", and are particularly in need of nurturing life. With population aging, changed household structure, the privatization of medical care, and uneven distribution of health resources, the need for older people to avoid chronicle illnesses, stay healthy, and slow down the process of aging is pressing. The enormous amount of self-health related knowledge, information, and advice put out or endorsed by the state apparatus, health authorities, and commercial interests to teach and promote how to live well and be healthy not only make participation of nurturing life practically necessary for older people, but also their moral obligation to the family and the society. Farquhar and Zhang (2012: 129) encountered many elderly Beijing residents who participated in *yangsheng* activities and talked about how "they looked after their own health so as not to become a burden either to their children or to the state" . For many commercial *paidu* providers, appropriating *yangsheng* language in advertising their products and services to the aging population is clearly a business move to tapping into the large *yangsheng* market.

However, once the concept of *paidu* is firmly entrenched in the *yangsheng* narrative, "the body" to which *paidu* attends is also undergoing reconstruction and made different from the imagined mechanical structure of physical body that needs to be periodically washed and cleaned. "The body" that *yangsheng* speaks of and attends to is rather the living body or the embodied person engaging in everyday practice that is concrete and meaningful—spatial and temporal as well as social and personal. For the urban residents who utilize *paidu* as a form of *yangsheng*, their intimate and mostly private bodily regime of *paidu* acquires a deeper connection to the cultural tradition and also a moral sense of participating in "cultivating, crafting, and nurturing the forms of life that are good, for themselves, their families, their communities." (Farquhar & Zhang 2012: 15)

Then, *paidu* is no longer intended for the care of the material body only, but also of "the heart" that is affective and mindful as *yangsheng* experts insist that nurturing life is ultimately nurturing the heart (养心).

Xueluo Huangcun wrote on 12 August 2010:

> Water gets in my heart? In the process of doing cupping *paidu*, I noticed that the spots that produced more blisters were the heart-related areas. The heart regulates emotions, and now I am more acutely aware that for women, suppressed emotions, depression and anxiety, and excessive stress could be causes of gynecological tumors.
>
> Six days after the second round of cupping treatment, my body was feeling relaxed gradually and more energetic, and my mood was getting better, too. This could be because I let water in my heart get out and I could feel relaxed and comfortable, again. I was amazed at this wonderful connection between the world and the mindful body. The emotional hurt can cause illnesses, and draining the heart water (心水) out of the body can also help one feel much better in the mind, emotion, and in the body.

She continued on 7 October 2010:

> I put cups on wherever is itchy on my body. My body is very smart and she (*ta*-female) often alerts me where I have problems (this

is because for more than three years I haven't used any Western drugs or creams to hurt her, and she has gradually regained the functions of *paidu* and self-healing).

Xueluo Huangcun's reflection on her *paidu* experience offers a sober reflection of personal struggle with health conditions and also a compelling story of using *paidu* to gain the sense of agency in transforming embodied experience and making life better. Her *paidu* experience is not typical part of *yangsheng* stories frequently displayed in the urban space along the alleys and streets or in the neighborhood parks. Then, *paidu* does bring something new to *yangsheng* movement. Conventional *yangsheng* practices tend to operate on the positive principle of activating, strengthening, and boosting *zheng qi* (right/ defensive *qi*) to fend off pathogenic invasions, emphasize "the wisdom of not getting sick," [39] and are often associated with various activities that are fun and enjoyable such as dancing, singing, exercising, playing chess, or simply chatting and walking in small or large groups. Whereas, *paidu* as *yangsheng* adds a different dimension: what to do or how to nurture life back to normal when "water gets into the heart" , or how to effectively rid the heart and the body of the burden of negative accumulations and transform the *du*-afflicted emotions and experience.

As *Xueluo Huangcun* shows, *paidu* is "philosophized" as "nurturing life" , a "technique of body-person" embraced by the urban masses, endorsed by the healthcare experts, marketed by the health industry, and actively promoted by the state. It is effective, mostly at the micro-level in the private space for the purpose of "*du shan qi shen*" (keeping oneself good and healthy), which can be interpreted as a creative adaption to the change of national healthcare policy from "public health" to "self-health" (see also Farquhar & Zhang 2012: 14–15). However, even in contemporary transformed "deep China", the formation of "self" still requires the participation of others, and this Chinese role-bearing self [40] is expected to grow, extending outward to the family, the community, and the society, as exemplified by Xueluo Huangcun, who practices *paidu yangsheng* as self-care, but also performed *paidu* on her parents and her daughter and shared the experience and knowledge gained through the practice with her online community.

Conclusion

Paidu therapeutics first emerged as manufactured commodities of Chinese herbal medicine and flourished in the consumer markets in the late 1990s. The concept and the practice were then contested and reinvented as a theoretical and therapeutic modality of traditional Chinese medicine and routinely incorporated in the TCM clinical work for treating ordinary illnesses and sub-health conditions. Recently *paidu* therapeutics have been promoted as forms of *yangsheng* or life nurturance. This trajectory of the *paidu* movement in contemporary China reflects ongoing negotiation and interaction among various healthcare needs of diverse populations and the interests of the health industry for growth and profits, and also the shifts in public health policy and the biopolitics at the local and national level in contemporary China.

The emergence and development of *paidu* mixture as Chinese medical preventive care also reflects the global influence in remaking traditional Chinese medicine and forming new patterns of producing and consuming Chinese medical knowledge and materials.

Chinese medicine has been a collaborator, playing a crucial role in the rise of *paidu* as a preventive health practice, a medical treatment, a health fashion, a life-style, and a mode of self-making and caring. This also suggests there is always a need and a motivation for Chinese medicine to search for ways to constantly renew itself in response to social, historical, economical, environmental, and political changes of the time. This could be true two thousand years ago when the *shanghan xuepai* 伤寒学派 rose, or 500 hundred years ago when the *wenbu xuepai* 温补学派 rose, or 300 hundreds years ago when the *wenbing xuepai* 温病学派 rose.

Yet, the unprecedented amount of exchanges of the global ideas, capitals, and technologies make new healthcare inventions, like *paidu*, a real mixture, which is fluid in identity and meaning and multifunctional in practice. A question to ponder could be—what will happen to Chinese medicine in the globalized future? Will its future involve a profound shift in emphasis from medicine as knowledge and practice in "pursuing the goal of facilitating life" [41] to "medicine as an economic activity?" (Scheid 2002: 151). An interesting example is Guangzhou based Clifford Hospital or Qi Fu Yiyuan in Chinese, a large modern general integrative hospital that provides medical services in Chinese medicine, Western biomedicine, and naturopathic medicine. It was founded in 2001, and became the first hospital in China accredited by the Joint

Commission International (JCI) and was also approved as an A-class hospital of traditional Chinese medicine (三甲中医医院) by the State Administration of Traditional Chinese Medicine of PRC. On the website of the hospital, one can find *Zhi Weibing Ke* 治未病科 (department of treating not-yet ill), and *Paidu Yangsheng* is one of the featured therapeutic programs in the department, which is a real mixture in language, concepts, and therapeutic methods including chelation therapy, traditional Chinese medicine, Western medicine, and nutritional therapy. [42] As displayed on the website of the hospital, the philosophy of care is to provide prompt, affordable, and appropriate treatment, and make the experience of healthcare enjoyable. Could this be the globalized future of traditional Chinese medicine? It could be one of many on a full spectrum of exchanges and cross-breeds between Chinese medicine and other medical knowledge, technologies, and practices.

Notes

[1] *Paidu* is often translated as "detox" or "detoxification". It is very likely that the idea of *paidu* was introduced or influenced by various practices of "detox" outside China. Yet, both "*pai*" and "*du*" are commonly used Chinese words, and the concept of *paidu* itself, being networked into Chinese medicine, has acquired distinctive meanings from the context of TCM. Therefore I choose to use *paidu* to avoid making a conceptual equivalence between the Chinese term of *paidu* and the English word "detox".

[2] See the conference announcement at the website of the China Association of Chinese Medicine: <http: //www. cacm. org. cn/coboportal/portal/channel_tzgg. ptview?funcid=showContent&infoLinkId=11488&infoSortId=52281>, accessed January 2011.

[3] See Anthony Giddens' discussion of tradition in *Runaway World* (2003: 36–50).

[4] See the introduction to the Center for Chinese Medical Holistic *Paidu* at Guangzhou Zijin Hospital, <http: //www. zijing120. com>, accessed January 2011.

[5] See Lang (2008). One of the five case studies in the book *Chanyelian Yinmou II: Yi Chang Meiyou Xiaoyan de Zhanzheng* (*Manufacturer's Chain of Conspiracy II: A War without Smoke*)is about the health product industry (*baojian pin chanye)* in contemporary China.

[6] According to Lang (2008), for a decade, Panlong Yunhai's *paidu* capsule has been leading the sales of the whole health/dietary products and sales totaled 5 billion Chinese yuan.

[7] Ibid.

[8] The same question was used as a topic for a session in CCTV-10's popular TV interview show "*Jiankang Zhi Lu*" (The Way to Health) broadcasted on 11 September 2012.

[9] Judith Farquhar, in the chapter "Medicinal Meals" of her book *Appetites* (2002: 48–55), offers a detailed discussion of Chinese medical technique of *bu* with meanings ranging from repairing, supplementing and bolstering to nourishing.

[10] In the high time of socialism, consuming herbal tonics to bolster personal health at the level of individual or private family might exist, but was unlikely promoted openly for public health in general. Instead, Chinese medical knowledge was frequently incorporated into the collective health projects as a sort of low-cost technique to provide basic healthcare and to prevent the spread of infectious diseases. Some of the socialist healthcare legacies extended well to the present. For example, the Eye Massage Exercise (*yan baojian cao*) created in 1963, based on the knowledge of the meridian systems and acupuncture points, is still part of the daily preventive exercise routines in elementary and middle schools. As late as the end of the 1980s, some work units still distributed bags of medicinal tea to their employees during a flu season to strengthen the collective resistance to the disease.

[11] For a detailed and informative discussion of deficiency and excess in relation to Chinese medicine and food consumption before and after China's economic reform, see Farquhar (2002: 121–163).

[12] One initial motivation for using *paidu* products by females, notoriously by stay at home wives of rich business men, is mainly for treating skin problems and losing weight.

[13] See "Center for Natural *Paidu*" (*Ziran Liaofa Paidu Zhongxin)*. Anonymous. http: //zypd120. com/show_news. asp?id=126. Accessed January2011.

[14] See the official website of the Panlong Yunhai Pharmaceuticals. http: //www. panlongyunhai. com. cn/index. asp. Accessed January 2011.

[15] See "the Basic Knowledge of *Du*" (*du de jiben zhishi*). Anonymous. http: // zypd120. com/show_news. asp?id=83. Accessed January 2011.

[16] See "A Different *Paidu* Therapy" (*bu yiyang de paidu fangfa*). Anonymous. http: //www. zypd120. com/show_news. asp?id=106. Accessed January 2011.

[17] Ibid.

[18] Duthie (2005: 1).

[19] Csordas (1994: 2).

[20] Until recently, many Chinese herb-based medicinal products were officially classified as health product (*baojian pin*) rather than medicine. The Yunnan-based *paidu* capsule was licensed as Chinese medicine (*zhongyao*) based on the new regulations passed in 2001.

[21] See Yu (2007); Zhao (2007).

[22] See the official website of the Panlong Yunhai Pharmaceuticals. http: //www. panlongyunhai. com. cn/index. asp. Accessed January 2016.

[23] See the article "*Paidu* drugs should not be used freely (*Paidu yaowu bu ke suibian shiyong*," anonymous, http: //www. 39. net/fitness/jfztl/pdzt/ywpd/84505. html.

[24] See 说文解字 .

[25] See Yu (2007: 7).

[26] See "A different type of *Paidu*" (*Bu yi yang de paidu*), anonymous. http : // zypd120. com/show_news. asp?id=106. Accessed January 2011.

[27] See Yu (2007: 7).

[28] See Zhao, 2007: 14–16.

[29] See Yu (2007: 183–184).

[30] For an informative discussion on treating female infertility in Chinese medicine, see Farquhar (1994).

[31] This patient was a friend whose paternal grandfather was a well-known local Chinese medical doctor. I had numerous casual discussions with her about health related topics. When I asked if I could use our conversations for my research, she not only gave me her full consent, but also emailed me her herbal prescriptions.

[32] See Yu (2000) and Zhao (2007). See also Jiang, Liangduo et al. (2003) for the discussion on *paidu* and the brain health.

[33] See http: //www. zijing120. com, last accessed 2 January 2011.

[34] See the commentary "*Paidu* is also a form of *yangsheng: a* new concept of health" in *People's Daily* (09/14/2012), which enlists Professor Jiang Liangduo, a well-known TCM scholar-physician at the Dongzhimen Hospital for expert opinions. Jiang has been an advocate for *paidu* as Chinese medical *yangsheng* that helps "keeping 'bodily channels' open and clear" (*baochi renti 'guandao' changtong*). See http: //news. xinhuanet. com/health/2017–09/19/ c_1121683334. html.

[35] These *episodes* broadcasted on 6/18 and 6/19, 2012 can be accessed online, see https: //www. youtube. com/watch?v=GLCpdFJRSj0; https: //www. youtube. com/watch?v=hzNj2c60tnQ.

[36] Farquhar and Zhang's recent book *Ten Thousand Things: Nurturing Life in Contemporary Beijing* (2012) shows diverse activities and techniques that fall into the purview of modern *yangsheng*. Some are clearly influenced by the global development in science, medicine, and psychology.

[37] See the article "*shenti da saochu: paidu yangyan jiaonang hao bangshou* (bodily cleaning: *paidu yangyan* capsule is a good assistant", http: //www. trends. com. cn/info/beauty/2014-01/501545. shtml.

[38] The case referred in this section is from the blog site http: //blog. sina. com. cn/snowingfield, which was last accessed by the author in January 2011. The original text is in Chinese, and the translation is by the author.

[39] "The wisdom of not getting sick" (*bu shengbing de zhihui*) is a title of a popular self-health *yangsheng* book by a famous folk Chinese medicine practitioner, Ma, Yueling, who is also hailed as her followers as *Jiankang Jiaomu* (Godmother of Health).

[40] See the discussion of "Confucian role ethics" in Ames (2011).

[41] See Lu, Guangshen et al (2007: 201). Lu Guangshen defines traditional Chinese medicine as "循生生之道，助生生之气，用生生之具，谋生生之效".

[42] See http: //www. clifford-hospital. org/zhiwei/index. html Accessed January 2011.

References

Ames, Roger.

2011. *Confucian Role Ethics: A Vocabulary.* Honolulu: University of Hawaii Press.

Chinese National Commission on Approving Terms of Science and Technology.

2004. *Chinese Terms in Traditional Chinese Medicine and Pharmacy.* Beijing: Science Press.

Csordas, Thomas.

1994. *Embodiment and Experience.* Cambridge: Cambridge University Press.

De Certeau, Michel.

1988. *The Practice of Everyday Life.* Berkeley: University of California Press.

Duthie, Laurie.

2005. "White Collars with Chinese Characteristics: Global Capitalism and the Formation of a Social Identity." *Anthropology of Work Review*26 (3): 1-12.

Etkin, L. Nina.

2006. *Edible Medicine: An Ethnopharmacology of Food.* Tuscon: The University of Arizona Press.

Farquhar, Judith.

1994. "Objects, Processes, and Female Infertility in Chinese Medicine." *Medical Anthropology Quarterly,* 5 (4): 370–399.

Farquhar, Judith.

2002. *Appetites: Food and Sex in Post-Socialist China.* Durham: Duke University Press.

Farquhar, Judith and Zhang, Qicheng.

2012. *Ten Thousand Things: Nurturing Life in Contemporary Beijing.* New York: Zone Books.

Giddens, Anthony.

2003. *Runaway World.* New York: Routledge.

Goldschmidt, Asaf.

2009. *The Evolution of Chinese Medicine: Song Dynasty, 960–1200.* New York: Routledge.

Good, Byron J.

1994. *Medicine, Rationality, and Experience.* Cambridge: Cambridge University Press.

Fassin, Didier.

2007. *When Bodies Remember: Experiences and Politics of AIDS in South Africa.* Translated by Amy Jacobs and Gabrielle Varro. Berkeley: University of California Press.

Hanson, Marta.

2006. "Northern purgatives and southern restoratives: Ming medical regionalism" in *Asian Medicine* 2 (2): 115–170.

Jiang, Liangduo, Zhou Lizhen, and Ma Yuan.

2004. "*Lun paidu jiedu, tiaobu yangsheng yu nao baojian*" (On detoxification, replenishing, and the brain health), *Chinese Journal of Information on Traditional Chinese Medicine* 11 (2): 100–101.

Lang, Larry.

2008. *Chanyelian Yinmou II: Yi Chang Meiyou Xiaoyan de Zhangzheng* (Manufacturer's Chain of Conspiracy II: A War without Smoke). Beijing: Dongfang Chubanshe.

Latour, Bruno.

1993. *We Have Never Been Modern.* Trans. by Catherine Porter. Cambridge: Harvard University Press.

Latour, Bruno.

2005. *Reassembling the Social: An Introduction to Actor-Network-Theory.*

Oxford: Oxford University Press.

Lu, Guangshen, Fu Jinghua, et al.

2007. *Zai Bei-Da Xue Zhongyi* (Learning Chinese Medicine at Peking University). Beijing: China City Press.

Mattingly, Cheryl and Linda C. Garro.

2000. *Narrative and the Cultural Construction of Illness and Healing*. Berkeley: University of California Press.

Morris, David B.

1998. *Illness and Culture in the Postmodern Age*. Berkeley: University of California Press.

Ren, Yingqiu, et al.

1986. *Zhongyi Gejia Xueshuo*. (Theoretical Schools of Traditional Chinese Medicine) Shanghai: Shanghai Science and Technology Press.

Rose, Nikolas.

2007. *The Politics of Life Itself: Biomedicine, Power, and Subjectivity in the Twenty-First Century*. Princeton: Princeton University.

Samudra, Jaida K.

2008. "Memory in our body: thick participation and translation of kinesthetic experience." *American Ethnologist* 35(4): 665–681.

Scambler, Gramham and Paul Higgs.

1998. *Modernity, Medicine and Health*. London and New York: Routledge.

Voler, Scheid.

2002. "Remodeling the arsenal of Chinese medicine: shared pasts, alternative future" in *Annals of the American Academy of Political and Social Science* 583 (1): 136–159.

Yu, Junsheng.

2000. *Duxie Xueshuo yu Linchuang* (Chinese Medical Theories of Toxicity and Clinical Practice). Beijing: Zhongguo Zhong Yiyao Chubanshe.

Yu, Zhimin.

2007. *Zhong Yi Yao zhi Du* (Concept of Toxin/Poison in Traditional Chinese Medicine) ed. Beijing: Zhongguo Kexue Jishu Wenxian Chubanshe.

Zhang, Yanhua.

2007. *Transforming Emotions with Chinese Medicine: An Ethnographic Account From Contemporary China*. New York: State University of New York Press.

Zhao, Zhiqiang.

2007. *Zhongyi Duxie Xueshuo yu Yinan Bing Zhiliao* (Chinese Medical Theories of Toxicity and Treating Difficult Illness Cases). Beijing: Renmin Weisheng Chubanshe.

Promises and Perils of *Guan*: Mental Health Care and the Rise of Biopolitical Paternalism in Post-Socialist China[1]

Zhiying Ma*

Abstract

Over the last three decades, most psychiatric inpatients in China have been hospitalized against their will, by their families. The first national Mental Health Law, effective since 2013, has reinforced the family's rights and responsibilities in psychiatric care. The family's involvement is inscribed in the keyword *guan* (管), a polysemous word that can refer to caring about and being responsible for another individual, and/or to managing, governing and controlling interpersonal situations. Drawing on 32 months of fieldwork, my research examines the family's involvement in psychiatry, and explores the implications for the ethics, affects, and political economy of care and population

* Zhiying Ma, School of Social Service Administration, University of Chicago.

governance in post-socialist China.

In this paper, I will trace the circulation of a language and practice of *guan* between legal, psychiatric, and familial realms. I argue that a biopolitical paternalism has emerged in post-socialist China that demands and legitimizes the family's involvement in psychiatric care, particularly by invoking and reconfiguring the family's role in performing *guan*. This ideological and practical formation constitutes mentally ill patients as subjects of perpetual risk management. The cultural ethics of *guan* paternalism lends ideological legitimacy to the post-socialist state's population control policies. Meanwhile, through the recursive circulation of paternalist modes of governmentality such as *guan*, the actual work of care and control are relegated to families. This form of biopolitical paternalism thus produces shared vulnerabilities and ethical unease within families, as well as aggravating health disparities across the mentally ill population. I will conclude the paper by considering the conceptual efficacy and practical implications of biopolitical paternalism.

Configuring the Family in Chinese Psychiatry

If we spend any time in a psychiatric hospital in China, we will likely be struck by the fact that most of the inpatients have been hospitalized against or regardless of their will, usually by their family members. According to a conservative estimate of the early 2000s, involuntary admission by families accounted for 60% of all psychiatric inpatients, and involuntary admission by police or other public sector agents accounted for another 20% (Pan, Xie, and Zheng 2003). In China, although families have long been involved in the care of mentally ill patients—or, before the advent of psychiatry, people known as insane, crazy, or mad—the *ways* in which they are involved are historically variable. It is only since the market reforms of the 1980s that families have gradually come to be the central agent in securing biomedical and institutional treatment for patients. In the past decade, concerned activists in China have launched a vehement attack on the prevalence of involuntary hospitalization and on families' involvement in it. Despite such contingencies and challenges, the first-ever national Mental Health Law in China, effective since May 2013, has reinforced the rights and responsibilities of families in patient care and management. Under the law, psychiatric patients are automatically subjected to the guardianship of their family members, listed in order as spouses, parents, adult children, and other close relatives. The law grants family members the right to consent to patients' treatment and to decide the involuntary commitment of those whom they believe pose an actual or potential danger to themselves. The law also stipulates that families have responsibilities to provide for, look after, and rehabilitate the patients (NPC 2012).

Why does the family occupy such a critical role in psychiatric care in China, especially today? How do medical techniques and institutional policies configure the role of the family in psychiatric services? What can these configurations reveal about the affects, ethics, and political economy of medical care and population governance in contemporary China? These are the questions that drive my project. Note that "configure" here means both to represent by an image and to fashion or put together in a certain form, because how these forces represent the contemporary Chinese family also shapes how they interact with, intervene into, and regulate it.

To address these questions, I conducted 32 months of fieldwork between 2008 and 2014, in a variety of settings that were actively engaged in serving, monitoring, or challenging family engagement in psychiatric care. They included psychiatric hospitals, community mental health teams, community social work centers, family support groups, and human rights agencies, mostly in Guangdong Province. My interviews explored how the staff, patients, and family members at these settings conceptualized the role of the family, including its rights and responsibilities. I observed how people occupied such roles or conveyed their role expectations to each other in interactions, justifying their actions accordingly.

Note that Guangdong is a relatively prosperous province, though with sharp internal inequalities, and that its mental health infrastructure is more established than many other parts of the country. Doing fieldwork there allowed me to observe and analyze Chinese psychiatry in its optimal form. Meanwhile, in order to gain a more balanced view of different institutional and socioeconomic conditions, over the years I have conducted shorter field trips to various psychiatric hospitals, community mental health teams, and, where they exist, social work/rehabilitation centers in other parts of China, from metropolises like Beijing and Shanghai to small cities and rural counties in Southwest China. Moreover, in order to explore representations of the family in historical and contemporary debates related to Chinese psychiatry, I interviewed lawmakers, attended national conferences, and conducted archival and media analysis.

The patient population that I focused on was those diagnosed with serious mental illnesses. In the Chinese context, the term "serious mental illness" (*zhongxing jingshen jibing*) is an administrative category covering schizophrenia, bipolar disorder, schizoaffective disorder, paranoid disorder, epilepsy with psychosis, and mental retardation with psychosis (Ministry of Health 2012). Official statistics estimated that as of 2011, there were more than 16 million people with serious mental illnesses in China (Xinhua News Agency 2011). Most of the patients in the institutions I visited were diagnosed with either schizophrenia or bipolar disorder. Because of my interests in institutional configurations of family relations, I paid less attention to the fine distinctions between these illnesses or to the "accuracy" of the diagnoses; rather, I focused on the social experience these patients and their families held in common.

Guan and the Rise of Biopolitical Paternalism

A prominent way of understanding the family's role in relation to the seriously mentally ill patient is that of 管 /*guan*. *Guan* is written as one Chinese character. In Chinese, a single character often constitutes a word in and of itself. Most single-character words are polysemic; that is, a word has two or more somewhat related meanings, and only the context in which it is uttered can specify its meaning-in-use. Single characters can also be combined with others to make less ambiguous compound words. Therefore, depending on the context and the word combination, *guan* can refer to what English speakers might call concerning oneself with and being responsible for another individual, even caring about or for someone, like a parent; and/or it can reference to managing, governing, intervention, and control — like a police officer.

Here is an example that illustrates the everyday practice of *guan* and its intricacies: One day in the Fall of 2013, I accompanied Mrs. Dong, a woman in her late fifties, to visit her daughter Tingting on a locked psychiatric ward in Nanhua[2]. Two months earlier, when Tingting had become distraught and insomniac by chaotic experiences at work and in love, Mrs. Dong had taken her to the hospital, telling her that the visit would just be for a brief check-up. But Tingting was admitted and had been kept on the ward ever since. Now that Tingting's condition had been stabilized by medications, Mrs. Dong felt the need to plan for Tingting's life after discharge. As Mrs. Dong saw it, Tingting's workplace had proven to be too stressful an environment. In fact, any job that required Tingting to work "outside" on her own would probably expose her to undue stress or unhealthy love relationships. It would also make it impossible for Mrs. Dong to monitor her medications. Therefore, without Tingting's knowledge, Mrs. Dong had sent a resignation letter to Tingting's company, and had bought a small storefront near their home, expecting that she and her daughter could run an herbal tea stall together. On the ward, Mrs. Dong presented her plan to Tingting with a big smile on her face, saying, "From now on life will be more relaxing for you." "NO!" Tingting screamed, "I'm 30 years old. I'm not a kid anymore. Why do you still want to control (*guan*) me?" "You're sick, Tingting," sighed Mrs. Dong, "how can I not look after/care for (*guan*) you?"

Here, the same actions—the mother hospitalizing the daughter against her will, planning her future, protecting her from potential harm, and ensuring her medical compliance—are seen by the daughter as control, and by the mother as care. But their different ethical evaluations are contained within the same Chinese character, *guan*. We can thus see that the polysemy of *guan* allows for struggles over its practical meanings, ethical orientations, and relational implications.

Guan is not only invoked in everyday family strife. Psychiatrists also use the language of *guan* as they teach family members to monitor patients' symptoms and pharmaceutical compliance. Moreover, *guan* has dominated the legal/policy texts produced and actively promoted by the central government. The new Mental Health Law highlights *guan* as a principle of mental health work, but here *guan* takes on a more specific meaning as management (*guanli*). Interestingly, while the law opens by requiring "all facets of society" to participate in *guan* or comprehensive management of mentally ill patients, it quickly relegates almost all of this responsibility to the patients' families. Article 21 of the law requires: "If it appears that a family member may have a mental disorder, other family members shall help them obtain prompt medical care, provide for their daily needs, and assume responsibility for their *supervision* and *management*." (NPC 2012)

In this paper, I will trace the circulation of *guan* between family practices, psychiatric encounters, and legal or policy reforms. I am interested in how people define, practice, and evaluate *guan* in everyday family life, and how such actions and meanings are transformed when psychiatric and legal discourses come to emphasize but also reconfigure the family's role as the site and agent of *guan*. *Guan* is useful to think with, not only because it permeates mental healthcare in China—and for that matter, many other aspects of contemporary Chinese life—but also because it helps us think through the relationship between care and biopower. As Michel Foucault has famously argued, modern nation states have developed in tandem with the rise of biopower, that is, the set of knowledges, experts, and institutions that bring "life and its mechanisms into the realm of explicit [political] calculations" (Foucault 1978, 143). Chinese psychiatry is a mechanism of power, more specifically biopower, for it brings mental illness, especially the risks it poses to the patient or the public (Castel 1991, Rose 2010), into the purview of political calculations. However, existing literature on biopower has largely focused on the involvement of expert knowledge and institutional mechanisms in the disciplining of the individual body or regulation of the population—the latter function is also called biopolitics (Foucault 1978). It tends to ignore how biopower in general or risk management in particular is mediated by modalities of intimate practice (Taylor 2012), or in other words, how the subject of biopower is produced through everyday social relations.

Meanwhile, the workings of psychiatry also hinge upon an ethics of care, because in seeking treatment for the patient, family members assume responsibilities for vulnerable others (Levinas 1988) and explore visions of the good life (Mol, Moser, and Pols 2010). The literature on care has shed light on people's lived moral experience and strivings in the face of suffering (Kleinman 2009, Buch 2013, Mattingly 2014, Mol 2008, Stevenson 2014, Throop 2010), as well as the conditions of possibility and impossibility for mutual caring (Biehl 2005, Scheper-Hughes 1993). However, this literature sometimes tends to idealize care, treating it as spontaneously springing from loving family feelings (Shakespeare 2006). My research looks at *guan*, a locally specific practice of intimate relation, a care-like relation if you will, that has political import. It illustrates how biopower requires and transforms intimate practices of care, as well as the power relations and political potentials in the provision of care. Ultimately, it will help us denaturalize the modern notion of care.

As we saw in Tingting's case, *guan*, at least for the family member who performs it, means caring for the vulnerable other by intervening into and making decisions for her life, because "mother (or father, or spouse) knows best". Therefore, examining the enactment and circulation of *guan* can provide insights into both the ethics and the politics of paternalism. Discussions of paternalism are often found in the literature of medical ethics. There, paternalism is defined as "the interference with a person's liberty of action justified by reasons referring exclusively to the welfare... of the person being coerced". (Dworkin 1972) Grounded in the tradition of liberal individualism, this literature often takes the interests of the individual being subjected to paternalistic intervention for granted. It ignores how the assumed definition of the individual subject as a person with its own interests is shaped by sociopolitical conditions, and too often cannot see how it emerges in relation to different imaginations of the social order.

Pertinent to our discussion, social scientists have pointed out that socialism, as practiced in both the Soviet Union and Maoist China (1949–1978), ran on a model of state paternalism. The system cultivated citizens' identification with and dependence on the state, which was imagined as the source of life and the guarantor of needed resources (Borneman 2004, Verdery 1996, Walder 1988). Yet now that China has moved into the era of post-socialism, forces of marketization, privatization, and global capital are entangled with "the lingering effects of socialist institutions and practices". (Zhang 2001, 179) We have to ask to what extent the promises of state paternalism are still upheld, who actually assumes the authority and responsibility of being "paternal," and how political and scientific discourses constitute both the target and the agent of paternalistic practices. Also, if the notion of paternalism implies an omniscient and omnipotent authority, what does it look like to practice paternalistic governance in everyday life, especially by agents who are not so powerful? By tracking the circulation of *guan* in different realms and paying attention to its power effects, we can address these questions and understand how paternalism is enacted both as a medical ethic and a political ideology in contemporary China.

I have argued that in post-socialist China, a *guan*-style biopolitical paternalism has emerged. It constitutes mentally ill patients as subjects of perpetual risk management. The cultural ethics that approve paternalism lend ideological legitimacy to the post-socialist state's control over the population; meanwhile, through the circulation of paternalistic values from the state to the medical professional and then to the family, the actual responsibilities of care and control end up falling to families, particularly female caregivers. This biopolitical paternalism then produces relatively novel desires, vulnerabilities, and even forms of violence within families. To demonstrate my argument, I will consider the cultural ethics of *guan* as an idealized practice of parenting and socialization. I will then move on to analyze how this default meaning of *guan* is mobilized and molded by the psychiatric discourse and the state's mental health management, the latter including both the community mental health program and the Mental Health Law. Next I will discuss the power effects of this reconfiguration for both patients and their family members. I will conclude the paper by considering the conceptual and practical implications of biopolitical paternalism.

From Hopeful Parenting to Chronic Risk Management

Despite its polysemy, *guan* has a commonly established meaning for many Chinese speakers, that is, the ethical practice of parenting. As cultural psychologists and anthropologists have told us, when Chinese parents practice *guan* with their children, their seemingly stern behavior of control, discipline, and restraint is often accompanied by care, love, and sacrificial labor (Fong 2004, Xu 2017). Underlying these practices is an image of children as "weak, vulnerable, and dependent beings" (Saari 1990) who have to be protected and trained in an optimal environment by their more mature and knowledgeable parents. Parents engage in *guan* with the hope that their children can become fully human (*chengren*), acting in harmony with the social order (Chao 1994), and eventually no longer need *guan*. Because this *guan* seamlessly links individual development, parental aspiration, and social order together, scholars have argued that *guan* is "the characteristic feature of Chinese socialization" (Wu 1996).

This *guan* in parenting finds many parallels with popular approaches to *guan* for the mentally ill. Chinese people I have interviewed say that they commonly experience mental illness as a radical breakdown of the orderly world. This disorderly experience thus incites their desire for treatment (*zhi*) and management (*guan*), both of which signify a quest for proper order. In particular, because mental illness is seen as rendering patients extremely vulnerable, even returning adults to a childlike fragile state, family members—including those who are not parents, like spouses and siblings—often feel it necessary to take all matters of the patient's life into their own hands. Like Mrs. Dong, they take great pains to build a "stress-free" environment for the patients, to restrict the patients' potentially harmful behavior, and to retrain the patients in emotional management, social skills, and pharmaceutical compliance. Such practices of *guan* entail much sacrificial labor, especially on the part of female caregivers such as mothers, sisters, and wives. When Tingting was receiving treatment on the ward, Mrs. Dong spent a lot of time running around the city to check every livelihood option for Tingting, and she tirelessly brought her home-cooked meals every day. These efforts, Mrs. Dong hoped, would make Tingting become a happy and healthy human again.

It is with this desire for a restored order and a fully human family member that relatives first bring patients to the psychiatric hospital. Psychiatric advertisements and education often encourage or even incite this desire. For example, this ad headed by the name of a psychiatric hospital invites family members to see disturbing behavior in everyday life, however minor, as medical problems and to hospitalize their loved ones accordingly. Such advertising promises not only to bring a quick cure to the patient, but also to restore happiness to the entire family.

Figure 1. Bus Poster Advertisement for a Psychiatric Hospital

SOURCE: author

Is there someone like this around you?

- Someone who talks and laughs to himself?
- Someone who is suspicious and impulsive?
- Someone who is eccentric or dull?

Make the right choice.

Make your whole family happy.

Similarly, when Mrs. Dong first sent Tingting to the hospital, her doctor friend there promised her that medications would soon calm her daughter down. After two weeks of treatment with psychotropics, Tingting's symptoms did subside. However, the doctor now told Mrs. Dong not to celebrate too early. He pointed to Tingting's occasional sleeplessness and expressions of frustration as signs of illness fluctuation, indicating the risk of relapse. Because of the high risk of relapse, he said, Tingting would have to keep taking the medications after discharge for anywhere from two years to a lifetime. He also admonished Mrs. Dong to closely monitor (*guan*) Tingting's pharmaceutical compliance, even when Tingting's condition seemed stable. Meanwhile, the doctor rejected other practices of *guan* suggested by Mrs. Dong. For example, when Mrs. Dong told the doctor her concern that the antipsychotics were making Tingting overweight and self-conscious, he chastised her for micromanaging, or engaging too much in *guan* (*guan taiduo*), and interfering in his treatment plan.

From this case, we can see how hospital psychiatry reconfigures intimate practices of *guan* into a form of biomedical risk management. While families seek help from psychiatry with the desire for a quick cure, psychiatry replies by inscribing patients into a chronic trajectory of remission, risk, and relapse. Because patients are seen to lack insight — that is, they do not know that they are mentally ill and need medical help — psychiatrists often recruit and prepare family members to *guan* them. Note that here, the only appropriate *guan* from the medical point of view becomes biomedical risk management. That is, family members should mobilize their authority to ensure patients' medication compliance, use their intimate knowledge and attentiveness to the patient to detect possible signs of relapse, and, if these signs do occur, use family funds to re-hospitalize the patient. Since mentally ill patients' illness trajectories are so often projected by psychiatrists to be chronic, family members' responsibilities of *guan*, in the form of biomedical risk management, also become chronic.

The healthcare structure in which *guan* operates is a relatively recent market formation. Compared to today, the socialist era before the early 1980s saw more agents involved in the care of mentally ill patients. Those agents included not just families, but also work units, neighborhood committees, the police, local governments, and so on. And the services that were provided were much more diverse, eclectic, and community-based (Kao 1979). However, since the Reform period began in the early 1980s, the community health care system in China has basically collapsed, and the public health insurance system has also been severely disrupted. Granted, the state has not completely withdrawn from healthcare provision, but its spending has been concentrated on building hospital infrastructures, especially large secondary or tertiary care hospital centers. These hospitals, in turn, run on a fee-for-service model. In the field of psychiatry, the rise of big hospitals has been accompanied by the dominance of a biomedical and pharmaceutical approach to mental health (Phillips 1998). Families, therefore, have come to desire hospital care, having been encouraged to see it as the only legitimate form of professional care for people with mental illnesses.

Because of the rollback of the welfare state, families are not only the main agents in hospitalizing patients, they are also the main payers of expensive hospital bills (Pearson 1995). In recent years, the state has been rebuilding its public health insurance system, but the system works mostly for people who are in the workforce (Blumenthal and Hsiao 2005). For unemployed psychiatric patients, their family members have difficulty signing them up for the meager insurance that is available; insurance often only covers inpatient stays, in any case. As a result, the mentally ill patient is constantly shuttled between the psychiatric hospital—typically a locked ward—and the home, with no end in sight. I call this structure of mental healthcare in the market reform era an "institution-family circuit" (Ma 2014). [3]

When psychiatrists tell family members to manage the patients and monitor their risks, the risk management is supposed to lower their chance of relapse and rehospitalization. However, in the hospital-family circuit, family members' heightened awareness of risks and their desire to *guan* sometimes lead to lowered thresholds for involuntary hospitalization. In the hospitals that I visited, more than half of the patients had been repeatedly hospitalized by their families. Some complained to me that while the first hospitalization was appropriate, given their chaotic feelings or behavior, subsequent commitments were based on increasingly minor issues, such as their refusal to take medications that had heavy side effects, or emotional reactions that were understandable given the circumstances. Without listening to their explanations, their family members—often with the help of hospital staff—forcibly or deceptively brought them back to the hospital. One patient put her frustration in this way: "I feel I am being suffocated by all the *guan* from my sister. Whenever I lift a finger, she'll say I have relapsed and send me back in here. I am just a puppet on a string, being pulled here and there by others. I am not able to resist the littlest bit."

Mental Health Policies and the Security State

If hospital psychiatry in the reform era has reconfigured familial *guan* into intimate practices that manage the patients' biomedical risk of illness occurrence and relapse, then the recent governmentmental health policies have further connected *guan* to security concerns and social management, highlighting another kind of risk. As Michael Dutton points out, the revolutionary people that was enshrined by the Maoist state has, in the post-socialist era, given way to the idea of a population that needs to be managed (Dutton 2005). Particularly, since the late 1990s and early 2000s, with the rise of socioeconomic inequalities and popular unrest, the state has been increasingly keen on maintaining social stability (*weiwen*) through developing a pervasive and dynamic social management system (Cho 2013, Lee and Zhang 2013). The new community mental health program, which was established by the Ministry of Health in 2004 and has been rolled out across the country in the last several years, capitalizes on this concern for security. While the program ambitiously claims to fill the large gap of care between the hospital and patients' homes, so far it has only targeted people diagnosed with serious mental illnesses. In fact, the scope of serious mental illness was by and large defined by community mental health policies. These disorders—schizophrenia, bipolar disorder, schizoaffective disorder, paranoid disorder, epilepsy with psychosis, and mental retardation with psychosis—almost all have a clear psychotic component, and they are presumed by community mental health program staff to make the patients prone to violence. [4]

Under the program, cadres of community mental health practitioners have been trained to manage (*guanli*) seriously mentally ill patients in the local community. My fieldwork with some of these practitioners shows that they are increasingly held accountable for incidents of violence committed by patients in their jurisdiction, and that they have learned to pass on the pressure they receive to manage patients to the families. Community mental health practitioners regularly visit patients' homes to collect information on their symptoms and pharmaceutical compliance, assessing the risk of violence. They do so not by talking with patients themselves, but by talking with their family members. The program has a humanitarian component, that is, it provides access to basic psychopharmaceuticals for impoverished patients. Yet community practitioners often dispense the medications into the hands of the patients' family members, so that they can supervise medication intake at home.

The program's official discourse celebrates open therapeutic alliances between patients and caregivers, and it denounces covert or coercive practices such as hidden pharmaceutical treatment, that is, mixing pills in patients' meals. In practice, however, community practitioners often acquiesce to such practices by families, and sometimes they even teach the families how to secretly medicate, with the aim of keeping patients' symptoms and risk of violence under check. Finally, home visits by community mental health practitioners often end with reminders to family members about the patients' riskiness and families' responsibility to watch over (*guan*) them. Therefore, in this emerging community mental health apparatus, patients with serious mental illnesses are further constituted as subjects carrying risks to public security and the social order. Correspondingly, *guan* is further reconfigured as security risk management. The responsibility of such management is passed down through layers of government to community mental health practitioners, and ultimately to the patients'.

The more recent mental health legislation process shows us how the reconfiguration and relegation processes have been legitimized. In the debates leading up to the legislation, patient rights activists vehemently criticized the extensive use of involuntary hospitalization in China. In response, the psychiatrists who drafted the law justified this practice as a manifestation of "state paternalism" or *guojia fuquan*, literally the power of the state to act as a father. Implicitly invoking the socialist legacy, they argued that this state paternalism had protected China from experiencing the disastrous consequences that deinstitutionalization had produced in the United States for both psychiatric patients and for public order (Xie and Ma 2011). This logic may sound persuasive, except that these psychiatrists could not get the post-socialist state to shoulder the concrete paternalistic responsibilities of *guan*. Instead, they invoked the notion of *guan* as a cultural ideal of kinship practice that could relegate responsibility for paternal care to families. One of them told me: "Fortunately, while American families can simply walk away, Chinese families will always *guan* their patients."

Prior to the legislation, families could hospitalize any member they deemed to be mentally ill as a form of "medical protection" (Shao et al. 2010). As a compromise with human rights campaigns, the Mental Health Law stipulates that admission to the psychiatric hospital should be voluntary in principle, except when patients pose risks to themselves or to others. Both activists and many patients were initially hopeful that this new principle would bring a reduction in constraints and an expansion of freedoms. On the other hand, many family caregivers found the law's emphasis on risks vague—indeed, the law does not define what counts as risk—eliding the many forms of vulnerability families experience when dealing with patients. In the cases I observed, a patient might have mood swings and curse his parents all the time, might be squandering the family's savings, or might simply wander off from home. While these behaviors deeply concerned the families, they did not necessarily qualify the patients for (involuntary) hospitalization—still the predominant treatment modality in many parts of China—depending on how one interprets the risk criterion in the law. In order to secure treatment for such patients, caregivers now have to tactically mobilize the law's language of risk, emphasizing the potentially grave consequences of such behaviors, sometimes even fabricating the patients' danger to the general public. Out of their biomedicalized sense of compassion and professional responsibilities, psychiatrists, community mental health practitioners, and other government agents often "collude" with caregivers in this process of strategic assessment. According to my observations, and reports from psychiatrists at different hospitals after the new Mental Health Law went into effect, after the initial few months of adjustment, involuntary hospitalization has again become the practical norm for hospital psychiatry. The only change that the law brought about seems to be that it has solidified the reconfiguration of *guan* as risk management in the minds of both family members and professionals.

Guan as Everyday Family Practice: Vulnerability, Indifference, and Unease

When the psychiatrists who drafted the Mental Health Law use the ideas of state paternalism and *guan* to legitimize involuntary hospitalization and other types of coercive care, they invoked the image of the state or the familial guardian as an omniscient, omnipotent, and ideologically masculine authority. Presumably, this authority figure always knows what's best for his subjects, always succeeds in achieving it, and will never harm or be harmed. Echoing the popular Chinese saying of "strict father, compassionate mother" , this authority figure also fulfills the cultural imagination of fatherhood, as he endows his subject with the essence of life and disciplines the subject according to normative principles of personhood. Granted, the mother also figures in the popular imaginary of familial *guan*, but she only adds an element of tender love, and does not disrupt the absolute power difference or the disciplinary goal implied in paternalism. But, what happens when this paternalistic ideal is practiced in everyday life?

First of all, because of the adult onset of many mental illnesses, the family members who engage in *guan* of patients usually do not wield much power. Most of them are aging parents or female relatives who have quit low-income jobs to stay at home and look after a patient. These vulnerable family members often lack the authority and even the physical strength to coerce the good behavior of a patient, so “how should I *guan*” is a question that preoccupies them every day. For example, in order to get the patients to take their meds, family caregivers whom I encountered often engage in trivial and yet painful negotiations with the patients. They either dole out “bribes” such as soft drinks or cigarettes, or mix the ground pills in meals while fearing discovery, or they take the pills themselves in front of the patients in order to prove that the pills are not poisonous. These soft tactics they devise for the hard task of *guan* are still met with silent aversion, outright resistance, or sometimes even physical confrontations by their patients. These reactions in turn can deeply upset family members, because they see the patients as misunderstanding and rejecting their most loving intentions. Such lack of appreciation could seem especially unfair, given that there are often other family members who wash their hands of *guan*. Tingting’s father, for example, spends his time playing mahjong, and avoids all the blame for her unhappiness.

Still, these emotional heartaches and physical exhaustion are nothing compared to caretakers’ fears of a precarious future. Because many adult patients don’t have spouses, children, or siblings, and because they are usually unemployed and receive little welfare subsidy, oftentimes it is their aging parents who need to “manage” them and use their meager retirement pensions to provide for them. Therefore, across the country, there is a widespread phenomenon perceived as “the old raising the disabled” (*lao yang can*)(SHDPESC 2014). These aging parents repeatedly ask this question to themselves, me (the researcher), and every professional or bureaucrat they encounter: when I become too old or pass away, who will *guan* or caringly manage my child?

Instead of reducing vulnerability in a patient population, *guan* reconfigured as biomedical and security risk management sometimes increases the vulnerability of those diagnosed as mentally ill. I have already mentioned how a heightened awareness of risks, inculcated by psychiatry, has led some family members to lower the bar for re-hospitalizing the patients in their care, and how this in turn has brought the patients a sense of suffocation. As Ian Hacking points out, how we categorize certain people influences how we interact with them and how they perform themselves, producing the "looping effects" that "make up people" (Hacking 1999). In this case, when patients' lives are disrupted by endless hospitalizations, restrictions, and familial strife, many of them do end up confirming gloomy psychiatric prognosis, which predicts that two-thirds of all patients with serious mental illnesses will be at least partially disabled throughout their lives. If *guan* as an idealized kinship practice is hinged on the production of hope, that is, the prospect that the subject of *guan* will become fully human again and not need to be subject to *guan* anymore, then what happens when *guan* in psychiatric practice systematically dashes such hope?

When hope becomes elusive, especially when patient's vulnerabilities make family ties more fragile, people feel uncertain about whether and how to take responsibility. [5] In many psychiatric hospitals that I visited, scores of patients had been left there for years, or even for the rest of their lives, by their family members. One such long-term patient, Xu

Wei, [6] became the plaintiff of the first and most sensational lawsuit under the new Mental Health Law. In May 2013, Xu filed suit against the hospital and his guardian, asking to be discharged. Diagnosed with schizophrenia, Wei had lived in a rundown hospital in Shanghai since 2003. He desperately wanted to be released, in order to start a family with the girlfriend he had met at the hospital. Everyone, including doctors at the hospital, agreed that Wei was stable and could function well outside the institution. Yet his elder brother, who was his guardian, refused to let him out, citing the petty fight between Xu Wei and their father that had resulted in his long-term commitment to hospital. The brother said: "I'm his guardian! I have to watch over (*guan*) him. I have to be responsible to society!" Wei suspected that his brother had ulterior motives, such as an unwillingness to share their now-deceased father's estate. In any case, the judge was initially sympathetic to Wei. He sent court officials to ask Wei's distant relatives and neighborhood committee whether they would like to be the guardian instead and endorse Wei's discharge. Nobody would.

In April 2015, Xu Wei lost the case. The court's verdict stated that as a patient with schizophrenia, Wei had limited legal capacity, and should be constantly managed (*guan*) with regard to his medications and everyday life. His guardian had both the responsibility and the right to arrange for such management. Given the family's circumstances, the verdict said, hospitalization was an appropriate mode of management (*guan*), and thus the guardian had *fulfilled* his responsibility by placing him there. [7] Xu and his supporters whom I interviewed found this decision ironic, because the problem was exactly that nobody wanted to *guan* Xu Wei, to concern themselves with his happiness and wellbeing.

Here we have a paradox of *guan*, a paradox that is conditioned by biopolitical paternalism. On the one hand, medical and administrative discourses constitute patients with serious mental illnesses as sources of chronic biomedical and security risks, and then relegate the entire responsibility for risk management to patients' families. So long as patients' risks are under control, where they are placed and how they are managed become a matter of legal and ethical indifference. In her recent ethnographic study of how the Canadian state handles tuberculosis and suicide epidemics among the Inuit, for example, Lisa Stevenson discusses an anonymous kind of state care, a biopolitical indifference. That is, so long as the physical lives of Inuit people are maintained, it does not matter to the state who it is that lives or dies, or how that life is lived (Stevenson 2014). This biopolitical indifference is also found in psychiatric care in China, except that rather than being mainly performed by the state, it is mediated by families. On the other hand, since *guan* reconfigures as risk management but also invokes an idealized kinship practice, this biopolitical indifference generates affective and ethical unease for people who receive or perceive it. After all, as a cultural ideal, *guan* hinges upon intimate affects and kin relations, requires attention to the vulnerable person's concrete circumstances, and hopes to produce differences in life. For Xu Wei and those who supported him, then, *guan* as an indifferent form of risk management betrays the spirit of hope and the play of difference also inherent in *guan*.

This indifference should not be blamed only on families that make the decision out of their particular circumstances, but also on the broader security state apparatus, which quietly but actively sought to maintain long-term hospitalization. In 2014, at a conference in Beijing, I spoke with a psychiatrist who was also an appointed mental health expert of the Ministry of Health. Because a few long-term inpatients had requested my help to get them out of the hospital, I asked the psychiatrist whether it might be possible for them to live outside together in a rental apartment, with regular visits by social workers—a dream of those inpatients. "No way," he responded firmly. "The government is concerned with protecting society rather than protecting the patient. If nothing bad happens with such group homes, that's fine. But if anything goes wrong, who carries the liability?" Frustrated, I asked him what one should/could do for those inmates, or for people like Xu Wei. "You can't openly help them," he said, "you might do something under the table. But I'm telling you, the law [about discharge] won't be loosened up."

Conclusion

In this paper, I have traced the circulation, reconfiguration, and transformation of *guan* through familial, psychiatric, and legal-administrative realms. I have shown the emergence of what I call "biopolitical paternalism" in mental health care in post-socialist China. This biopolitical paternalism has three features: 1) Formation of certain kinds of biological subjects and of a need for their regulation. In particular, the psychiatric, administrative, and legal discourses of Chinese mental health constitute mentally ill patients as subjects posing permanent biomedical risks to themselves, and embodying security risks to the public. The subjects constituted as such require constant management so that a healthy citizenry and good public order can be produced. 2) Ideological legitimation and structural displacement. Although *guan* has now been reconfigured as risk management, the cultural ethics of *guan* as an idealized kinship practice allows the state to claim a socialist legacy and fashion itself in the process of mental health management as a caring parent state. *Guan* paternalism also encourages the state to displace such management responsibilities to the families. 3) Power effects: production of intimate vulnerabilities and ethical unease, but also new political potentials. While the agent of paternalism is imagined to be an omniscient, omnipotent, and ideologically masculine authority, in reality it is often vulnerable family members, such as women and the elderly who must do the managing. The responsibility for chronic risk management strains caretakers' relationships with patients, making both sides even more vulnerable. In particular, the chronicity and indifference implied in this form of risk management generates deep ethical unease about *guan*. Meanwhile, in response to the tensions between the ideological legitimation and structural displacement of biopolitical paternalism, some family caregivers I interviewed have begun to "flip the script" (Carr 2010), demanding that the state take the responsibility of *guan*, to act as a proper parent. They ask the state to recognize the vulnerability of both the patients and themselves, and heal the injuries wrought by marketization. As such, they reframe *guan* to build what I call "paternalistic citizenship" .

Mental health care in contemporary China is only a special case of a more general biopolitical paternalism. Because of its striking use of coercion/constraint, and because of the often clear presence of the state, psychiatry allows "the larger social 'will' (to power, to social order) [to] show its teeth" (Lovell and Rhodes 2014). Yet some other mechanisms of governance in contemporary China also share the conditions of biopolitical paternalism, such as the reconfiguration of "the people" into a "population" that needs to be managed, the neoliberal devolution of welfare and healthcare, and the rise of the security state. For example, the famous one-child policy made married couples responsible for producing fewer children, but children thought to be of better "quality" , in order to save the nation-state from a "population crisis" (Anagnost 1995, Greenhalgh 2008). As the post-socialist state increasingly expands its social management, other non-familial relations, such as neighbors and school teachers, are pulled in to perform *guan* as well. Throughout the world, as biomedicalized techniques of rule continue to redefine individual wellbeing and population security, and as neoliberal economic policies continue to transfer responsibilities to achieve these new ideals away from the state, scholars have noticed the use of paternalistic strategies in diverse projects of social governance, from poverty alleviation to health promotion. These projects draw on everyday relations and techniques of direction, instruction, and surveillance to produce individual subjects who are either self-governed or well managed (Borovoy and Roberto 2015, Brodwin 2012, Soss, Fording, and Schram 2011). I am not suggesting that these formations of power are equivalent; far from it. My point is that as an analytical concept, "biopolitical paternalism" may help us detect, in different relations and governance mechanisms, how subjects are constituted and regulated, how responsibilities for discipline are legitimized and distributed in social space, and the power effects of such regulations, especially in the context of intimate relations.

An analysis of biopolitical paternalism and its intimate workings can also contribute to our understanding of chronic shortfalls in the provision of health care. In the case of mental health in China, during the mental health legislation debate, whenever critics raised concerns with the pervasiveness of involuntary hospitalization and its potential abuse, leading psychiatrists would argue that rather than having too much institutionalization, China didn't have enough. This statement is in one sense true. As of 2011, China had only 1 psychiatric bed in hospitals per 100,000 population, which was said to be just slightly better than the low and lower-middle income countries (0.4 and 0.6 respectively), but fell far behind the upper-middle income countries (2.7)—to which China as an economic entity belongs—and the high income countries (13.6) (WHO 2011a, b). However, it was not until very recently that those psychiatrists have begun to admit another, less convenient, fact: for Chinese patients who are sent to the hospital, their hospitalization is usually much longer, and their readmission rate is higher, compared to many other countries. [8] While no national statistics have been published yet, local reports have slowly emerged. In 2013, the director of Anding Hospital in Beijing reported that out of the 800 inpatients of the hospital, over 100 of them were long-term patients; the longest stay had been 25 years. In a survey done by the staff at another psychiatric hospital in Beijing, 180 inpatients out of the total 300 in the hospital wanted to go home, and among them 150 were considered to be in stable condition (Zhang 2013).

Therefore, the question is not simply to take sides between these two phenomena—fewer hospital beds on the one hand, longer hospitalization and higher readmission rates on the other—or between institutionalization and deinstitutionalization. Rather, we should see these two phenomena as emerging from a common landscape of inadequate mental health care, and understand how this landscape is shaped by the political economy and intimate anxieties generated in the course of psychiatric care in its present form. As I have pointed out, in the hospital-family circuit, families with means pull together either their own funds or welfare resources to put the patients in the hospital. But many of these families end up being trapped in the chronic trajectory of risk management punctuated by endless cycles of hospitalization. Meanwhile, families that don't have the means for hospital care receive little by way of professional services or relief. In order to address this disparity, we have to consider: if there really aren't enough psychiatric institutions in China, what kind of institutions do we need? Are we simply going to duplicate the existing total institutions that operate on, and instill, the logic of chronic risk management, and that rely on familial resources and popular triage to function? To answer these questions, we would need to look at patients and families who don't have ready access to psychiatric hospitals, especially in areas where professional psychiatry isn't established, in order to understand how those patients and families live, and what services they need.

On Sept. 27, 2017, after 15 years of hospitalization, Xu Wei was finally released. In July of that year, the Center for Forensic Science under the Ministry of Justice had issued a statement certifying Mr. Xu's "full legal capacity" , which meant that he could make decisions for himself. While celebrating his hard-won freedom, the activists who had been supporting him were hesitant to claim this as a victory. After all, they argued, legal capacity should be inherent in every human being; the fact that it was evaluated and granted by forensic psychiatrists only reinforced the medical authority. Moreover, at the same time as Xu's release, some major cities were issuing new policies to bring long-term hospitalization to patients whose families could not care for them, as an explicit attempt to "reduce risks to social control (*guankong*)" . Therefore, instead of an end to long-term hospitalization, we may be witnessing a slow and quiet change in the formation of biopolitical paternalism: when families fail, the increasingly wealthy state may now be more willing to step up and even more fully assume the role of the parent, directly securing the biopolitical order and deciding the fate of all its children.

Notes

[1] A slightly different version of this paper was published in *Medicine Anthropology Theory*, 2020, 7(3).

[2] Nanhua is a pseudonym, as are the names of most persons and organizations that I study. In China, psychiatric hospitals are few and far between, making them easily identifiable. It is even more so for non-medical organizations serving the mentally ill population, such as social work centers. Therefore, in order to protect my interlocutors, I choose to anonymize not only the organizations with which they are associated, but also the places of the organizations. When necessary, I will also change the identification details or report several persons/organizations/places as a single case.

[3] This term is inspired by, and stands in contradistinction with, Kim Hopper's analysis of the "institution circuit" for people with serious mental illnesses in the United States. The latter refers to "shelters and custodial facilities linked in haphazard chains of limited-time occupancy" (Hopper et al. 1997). Both the

institution-family circuit in China and the institution circuit in the U. S. are conditioned by the neoliberalization of social services, and they both produce "revolving doors" for patients/service users.

[4] In contrast, although the new millennium has witnessed the rise of depression as the most prevalent mental illness in China, the national community mental health program does not include it as a targeted illness. This is partly because, as the program's leaders told me, people with depression typically only harm themselves, not others, thereby not concerning the security state as much (Ma n. d.).

[5] Similarly, Sarah Pinto argues that in India, psychiatric care often mediates the dissolution of kinship bonds and conjugal love. As such, it "adds vulnerability to the already—and inherently—vulnerable condition of kinship" (Pinto 2014, 30).

[6] This is a pseudonym of the plaintiff that has been universally used by his attorney, other activists, and journalists.

[7] The verdict further stated that although Article 44 of the Mental Health Law granted voluntary inpatients the right to voluntary discharge, Xu had been involuntarily hospitalized because of his risk to others, and thus could not enjoy this right. Activists have criticized this provision, or the court's interpretation of it, for depriving people of their freedom indefinitely merely for a one-time behavior. Some have demanded a legal procedure for involuntarily hospitalized patients to transition to voluntary hospitalization, and thereby to be eligible for voluntary discharge.

[8] In the 2015 conference of the Society for Psychological Anthropology, held in Boston, Dr. Yu Xin, director of the Peking University Mental Health Institute, pointed to "longer hospitalization and higher readmission rate" as one of the biggest problems in China's mental health system.

References

Anagnost, Ann. 1995. "A surfeit of bodies: population and the rationality of the state in post-Mao China." In *Conceiving the New World Order: The Global Politics of Reproduction*, edited by Faye Ginsburg and Rayna Rapp, 22–41. Berkeley, CA: University of California Press.

Biehl, João. 2005. *Vita: Life in a Zone of Social Abandonment*. Berkeley, CA: University of California Press.

Blumenthal, David, and William Hsiao. 2005. "Privatization and its discontents—the evolving Chinese health care system." *New England Journal of Medicine* 353 (11): 1165–1170.

Borneman, John. 2004. "Introduction: Theorizing regime ends" . In *Death of the*

Father: An Anthropology of the End in Political Authority, edited by John Borneman, 1–32. New York, NY: Berghahn Books.

Borovoy, Amy, and Christina A Roberto. 2015. "Japanese and American public health approaches to preventing population weight gain: A role for paternalism?" *Social Science & Medicine* 143: 62–70.

Brodwin, Paul. 2012. *Everyday Ethics: Voices from the Front Line of Community Psychiatry.* Berkeley, CA: University of California Press.

Buch, Elana D. 2013. "Senses of care: Embodying inequality and sustaining personhood in the home care of older adults in Chicago" . *American Ethnologist* 40 (4): 637–650.

Carr, E Summerson. 2010. *Scripting Addiction: The Politics of Therapeutic Talk and American Sobriety.* Princeton, NJ: Princeton University Press.

Castel, Robert. 1991. "From dangerousness to risk" . In *The Foucault Effect: Studies in Governmentality*, edited by Graham Burchell, Colin Gordon and Peter Miller, 281–298. Chicago, IL: University of Chicago Press.

Chao, Ruth K. 1994. "Beyond parental control and authoritarian parenting style: Understanding Chinese parenting through the cultural notion of training" . *Child Development* 65 (4): 1111–1119.

Cho, Mun Young. 2013. *The Specter of "The People": Urban Poverty in Northeast China.* Ithaca, NY: Cornell University Press.

Dutton, Michael. 2005. *Policing Chinese Politics: A History.* Durham, HC: Duke University Press.

Dworkin, Gerald. 1972. "Paternalism" . *the Monist.* 56 (1): 64–84.

Fong, Vanessa L. 2004. *Only Hope: Coming of Age under China's One-Child Policy.* Palo Alto, CA: Stanford University Press.

Foucault, Michel. 1978. *The History of Sexuality, Volume I: The Will to Knowledge.* 1st American ed. New York: Pantheon Books.

Greenhalgh, Susan. 2008. *Just One Child: Science and Policy in Deng's China.* Berkeley, CA: University of California Press.

Hacking, Ian. 1999. "Madness: biological or constructed?" In *The Social Construction of What?*, edited by Ian Hacking, 100–124. Cambridge, MA: Harvard University Press.

Hopper, Kim, John Jost, Terri Hay, Susan Welber, and Gary Haugland. 1997. "Homeless, severe mental illness, and the institutional circuit" . *Psychiatric Services* 48 (5): 659–665.

Kao, John J. 1979. *Three Millennia of Chinese Psychiatry.* New York: Institute for Advanced Research in Asian Science and Medicine.

Kleinman, Arthur. 2009. "Caregiving: the odyssey of becoming more human" .

The Lancet 373 (9660): 292–293.

Lee, Ching Kwan, and Yonghong Zhang. 2013. "The Power of Instability: Unraveling the Microfoundations of Bargained Authoritarianism in China" . *American Journal of Sociology* 118 (6): 1475–1508.

Levinas, Emmanuel. 1988. "Useless suffering" . In *The Provocation of Levinas: Rethinking the Other*, edited by Robert Bernasconi and David Wood, 156–167. Routledge.

Lovell, Anne M, and Lorna A Rhodes. 2014. "Psychiatry with teeth: Notes on coercion and control in France and the United States" . *Culture, Medicine, and Psychiatry* 38 (4): 618–622.

Ma, Zhiying. 2014. "Intimate politics of life: the family subject of rights/responsibiliites and mental health legislation." *Thinking* 40 (3): 42–49.

Ma, Zhiying. 2020. "Numbers and the assembling of a community mental health infrastructure in post-socialist China." In *Can Science and Technology Save China? Utopian Dreams, Dystopian Realities*, edited by Susan Greenhalgh and Li Zhang. 25–49. Ithaca, NY: Cornell University Press.

Mattingly, Cheryl. 2014. *Moral Laboratories: Family Peril and the Struggle for a Good Life*. Berkeley, CA: University of California Press.

Ministry of Health, P. R. C. 2012. Rules for Management and Treatment of Serious Mental Illnesses.

Mol, Annemarie. 2008. *The Logic of Care: Health and the Problem of Patient Choice*: London, Routledge.

Mol, Annemarie, Ingunn Moser, and AJ Pols. 2010. *Care in Practice: On Tinkering in Clinics, Homes and Farms*. Transcript Verlag.

NPC, National People's Congress of the People's Republic of China. 2012. *Mental Health Law of the People's Republic of China*. Beijing: Law Press China.

Pan, Zhongde, Bin Xie, and Zhanpei Zheng. 2003. "A survey on psychiatric hospital admission and related factors in China." *Journal of Clinical Psychological Medicine* 13 (5): 270–274.

Pearson, Veronica. 1995. *Mental Health Care in China: State Policies, Professional Services and Family Responsibilities*. London: Gaskell.

Phillips, Michael. 1998. "The transformation of China's mental health services" . *The China Journal* (39): 1–36.

Pinto, Sarah. 2014. *Daughters of Parvati: Women and Madness in Contemporary India*. Philadelphia, PA: University of Pennsylvania Press.

Rose, Nikolas. 2010. " 'Screen and intervene' : Governing risky brains." *History of the Human Sciences* 23: 79–105.

Saari, Jon L. 1990. *Legacies of Childhood: Growing up Chinese in a Time of*

Crisis, 1890–1920. Cambridge, MA: Harvard University Asia Center.

Scheper-Hughes, Nancy. 1993. *Death without Weeping: The Violence of Everyday Life in Brazil*. Berkeley, CA: University of California Press.

Shakespeare, Tom. 2006. *Disability Rights and Wrongs*. Routledge.

Shao, Yang, Bin Xie, Mary-Jo Good, and Byron Good. 2010. "Current legislation on admission of mentally ill patients in China." *International Journal of Law and Psychiatry* 33 (1): 52.

SHDPESC, Shanghai Disabled People Employment Service Center, and East China Normal University School of Social Development. 2014. "Current situations of and thoughts on 'The old raising the disabled' : Shanghai as an example." *Disability Research* 1: 13–18.

Soss, Joe, Richard C Fording, and Sanford Schram. 2011. *Disciplining the Poor: Neoliberal Paternalism and the Persistent Power of Race*. Chicago, IL: University of Chicago Press.

Stevenson, Lisa. 2014. *Life Beside Itself: Imagining Care in the Canadian Arctic*. Berkeley, CA: University of California Press.

Taylor, Chloë. 2012. "Foucault and familial power." *Hypatia* 27 (1): 201–218.

Throop, C. Jason. 2010. *Suffering and Sentiment: Exploring the Vicissitudes of Experience and Pain in Yap*. Berkeley, CA: University of California Press.

Verdery, Katherine. 1996. *What Was Socialism, and What Comes Next?* Berkeley, CA: Cambridge University Press.

Walder, Andrew G. 1988. *Communist Neo-traditionalism: Work and Authority in Chinese Industry*. Berkeley, CA: University of California Press.

WHO, World Health Organization. 2011a. Mental Health Atlas 2011. Geneva, Switzerland.

WHO, World Health Organization. 2011b. Mental Health Atlas 2011—China. Geneva, Swizerland.

Wu, David YH. 1996. "Parental control: Psychocultural interpretations of Chinese patterns of socialization." In *Growing up the Chinese Way: Chinese Child and Adolescent Development*, edited by Sing Lau, 1–28. Hong Kong: the Chinese University Press.

Xie, Bin, and Hong Ma. 2011. "Six Myths Concerning the Mental Health Law." *Ma Hong's Blog*. Accessed June 15, 2015. http: //mahong2006. blog. hexun. com/65403321_d. html.

Xinhua News Agency. 2011. Director of Ministry of Health: 16 Million Patients with Severe Mental Disorders in China. Accessed Aug. 30, 2014. http: //politics. people. com. cn/GB/1027/15996111. html.

Xu, Jing. 2017. *The Good Child: Moral Development in a Chinese Preschool*.

Palo Alto, CA: Stanford University Press.

Zhang, Li. 2001. "Migration and privatization of space and power in late socialist China." *American Ethnologist* 28 (1): 179–205.

Zhang, Ran. 2013. "Many problems to be solved for the prevention and treatment of mental disorders." *Jinghua Times*, Apr. 18, 2013. Accessed Dec. 13, 2015. http: // news. ifeng. com/mainland/detail_2013_04/18/24337106_0. shtml.

中医如何成为只治慢性病的医学

艾理克*

对于外人来说，中医的有效性尚待考察，只有少数几个特殊的中医疗法经过了双盲临床试验，可以暂予认可。而中国国内的情况则更为复杂。尽管批判中医的大有人在，视之为缺乏临床疗效的迷信活动，但也有很多人认可中医的有效性及其治疗范围的广泛性。在中医院，我们随时可以看到类风湿性关节炎、银屑病、多囊卵巢综合征、糖尿病、胃食管返流、中风并发症、慢性乙型肝炎、不孕症、充血性心力衰竭、哮喘、头痛等等各色疾病的病人排着长队找中医专家治疗。据卫生部统计，2010 年有超过 32700 万门诊病人到中医类医疗场所就医。[1] 广州中医药大学附属第一医院自称是广州年就诊人数最多的医院，在 2013 年有 300 万的门诊和急诊量。[2] 但是，这种对中医临床的认可并非针对所有方面，患者和医生也同时认可西医在许多方面的疗效。关于中医临床的局限性，最广泛的共识就是它起效慢，适合治疗慢性病。这种认识基本上来自和西医的疗效对比，人们普遍认为后者见效快、长于治疗急症。确实也有这样一句老生常谈的谚语：“西医治急性病；

* 艾理克（Eric I. Karchmer），阿巴拉契亚州立大学（Appalachian State University）人类学系。

中医治慢性病。”

见效快的医学治疗急性病，见效慢的医学治疗慢性病，这种二分法貌似抓住了两种医学体系本质上的区别，一直十分盛行。1990 年代末，我在北京中医药大学的五年本科期间，基本没听说过有人对这个说法提出质疑。2008 年到 2009 年，我有幸采访了 39 位耄耋之年的老医生，才发现实际上民国时期（1911-1949）对这两种医学体系的认知并非如此。他们反复地告诉我在民国时期，中医治疗急性病的快速、有效是众所周知的。某些读者大概会对这些民国时代的认知感到吃惊。我作为一个中医大夫和研究者，第一次听到这些说法的时候也很是怀疑，后来随着进一步的研究和反思才逐渐接受。

民国距离现在并不遥远，我们该如何解释在不长的时间里，对中医疗效的公众认知的戏剧性变化？是 20 世纪早期的中国社会对西医的认识不够？或者患者对中医的熟悉使得他们对其过分信任，而后信任随着现代教育体系的传播而逐渐瓦解？本文强调民国时期的认识并非误解。访谈过 1949 年之前即开始行医的老医生之后，我认为有足够的证据支持民国时期的说法——中医师一度很善于用草药和其他治疗方法获得快速疗效。只是后来中医学本身和它所处的社会政治环境都改变了。今天，中医和西医都是高度体制化的国家医疗服务体系的一部分，1949 年前这体系并不存在。疗效“放慢”的中医是这种医疗体系的体制化过程带来的认识论改变的结果，而非从来如此。[3]

一、急诊室

当代中医的“见效慢”在争分夺秒的急诊医学领域是最明显的。与世界上的许多大城市一样，需要紧急医疗救护的病人大多就近在医院或医疗中心的急诊室得到处置。而且中医和西医医院都设有急诊室。

中国的医疗服务体系包含着多层级的公立诊所和医院，这些诊所和医院可以分为两套并行的医学机构，一边是中医一边是西医。尽管西医机构在如今占据优势，拥有大约75%—80%的医生、学校和医院，中医也还有一定的地位。[4]所有的大城市都有中医院，从1980年代开始，所有的县城也要求至少有一所中医院。除了推行两种不同的医学服务体系，国家政策还鼓励中西医结合。西医的综合医院只有设置中医科才能评上三甲，中医院的诊断和治疗也相当依赖西医和现代医疗技术。尽管政府支持中医机构，也有政策维持这两种主流医学实践的平衡，急诊行业明显倾向于施行西医的急诊医学救护措施。需要送急诊时，无论是中医院还是西医院的急诊科，病人的期待、医生的给予都是西医的急诊救护措施。

1990年代末，我在北京中医药大学附属医院实习的时候，亲历了急诊科对西医的偏向。所有的学生都要在急诊科做5周的实习。轮转到这里时，尽管早已知晓这个科室对西医的偏向，但中医学治疗手段在这里几乎尽付阙如，还是让我十分震惊。其他科室或多或少会强调西医，但中医总是必要的。学者和官员早已注意到了中医在急诊医学中的缺乏。1980年代早期，卫生部中医司就率先组织了中医内科急症进修班。曾于1950年代后期参加“西学中”培训的老医生黄星垣，受委托将相关的学习材料汇编成这一课程的首批教材之一。[5]类似这样的项目最终促成了1997年新教科书《中医急诊学》（第六版统编中医学教科书系列之一）的成书。这本新的教科书出版并被编入那一年的教学大纲时，我正好四年级，成为这本教材的第一批学生。尽管这门课的课时比中医内科、中医妇科、内科学、外科学等主课少，但老师们向我们反复强调它的重要性。而且许多附属医院的专家都被请来给我们讲课，不像其他课程那样通常由低年资的教师授课。

可惜的是，这门新的实验性课程的活力没有延续到我们的临床实习中。这或许并不奇怪，急诊科的医生基本没有参与这门课的教学。尽管除一位医生外，急诊科的医生们都接受过中医训练，[6] 但是他们在临床工作中几乎完全依赖西医的治疗手段，而非《中医急诊学》教科书中讲授的中医方法。在我实习时，临床带教老师正在钻研使用科里的新呼吸机。尽管她和她的同事们以自己掌握了西医学的急救技术和药物为傲，他们并不认为自己的临床工作抛弃了中医学，他们对中医还是有着很强的职业身份认同的。他们解释说只要患者的状况稳定了就可转到医院的其他科室接受更细致的中医治疗，他们还有充足的时间体验中医学的好处。

中医学在急诊室的缺失基本已经常规化，几乎无人质疑。我记得当时有一个非常理想主义的同学，他对这种现状十分不满。尽管现代的中医学教育包含了很多必修的西医课程，而且大多数医生和医学生都认可中西医结合模式，这个同学只想学习中医。他对课程设置对于西医的重视一直不满，而急诊科的实习将之激化成了一次危机，以至于有一天临床带教老师不得不专门把他叫到一边，跟他讨论了一下中医的“现状”。她解释说我们不能忽略西医学的优势，尤其是在急诊医学这样的领域，但是这个现实不会抹杀中医学的许多优点。在当今社会，纯中医是行不通的。她解释说，每种医学体系都有它的长处，我们不应该从思想上将自己限制于某一种体系。

我私下里跟我的老师讨论过能否在急诊医学里多用中医学治疗方法，她给了我一个比较实际的答案，强调了这个科室面临的供给障碍，尤其是草药制备方面。大多数处方通常包含十几味草药，这些药需要称量、配伍、熬制之后方能使用。即便有着熟练的药剂师的帮助，熬药还是需要相当长的时间，不少于 15 到 20 分钟，甚至一个多小时，不符合急诊对时间的严格要求。而且，医院药房和制剂室是为门诊和住院部服务的，没有相应的加急体系专门满足急诊科的用药需求。医院药房一般要花大约 1-2 小时来备好一个典型的门诊处方。住院患者的汤药耗时更长，因为必须要熬制好送到病房里。所以一个主治医生早上给住院病人开具的药方要到下午 3 点或更晚才能送到病人手中。情况更糟的是，医院的中药部每天下午 5 点就关门了，晚上急诊科大量患者入住时只有西药房是开着的。

尽管这些实际的障碍并非不能克服，它们的存在使得中医“慢”的感觉更明显了。一种解决方案是发展中药汤剂以外的剂型，这也是许多医生的观点。现在，易溶于水的中药颗粒剂很受欢迎，尤其是对于那些工作繁忙无法在家熬药的病人来说。1990 年代后期，我的一些临床老师已经开始试着使用颗粒剂，现在大部分的大型中医院的药房都提供中药饮片和颗粒。

对急诊医学和急救护理更有意义的另一剂型改革是中药注射剂。国家中医药管理局自 1990 年代开始确定急诊使用的必备中成药。中成药的不足在于它是预制的配方，医生无法根据病人的个体需求而调整处方，这恰恰是许多医生所强调的中医最大的特点。但是中成药便于立刻使用，适于大量生产，使得它们特别适用于急诊医学。因此，国家中医药管理局最先在 1992 年选定了 15 种急诊必备中成药，到 1995 年增加到 40 种，1997 年增至 57 种。[7]

在国家中医药管理局推荐的 57 种必备中成药中，我在急诊科实习期间常见的只有清开灵注射液，它基于名贵的安宫牛黄丸的配方制成，可用于急性中风或是高烧不退，医生们似乎还比较喜欢这个中成药，但也主要把它作为已有的西医治疗流程的补充。中医治疗在急诊室是如此少见，以至于我至今还清晰地记得，某天有个医生想用针灸治疗的时候，所有的实习生都跑来观摩。因为那是我们为期 5 周的急诊实习中唯一一次看到针灸治疗。病人中风后昏迷，呼吸困难，发热（发热或许是源自病灶不明感染）。首诊医生已经施行了一系列西医的抢救，一个住院医师决定试试针灸来为病人降温。他选择了病人十指的十宣穴施针，每个手指都挤出了几滴血。我们很兴奋地看着病人的体温表读数下降，他的体温在大约 15 分钟内降了 1 摄氏度。

二、民国时期的急性病

当年我也不知不觉地接受了中医见效慢、不适用于大部分急性病这个偏见。急诊科实习之后，我更觉得课本上的中医急诊实际在大多数时候并不能指导临床，只是一个理想的描绘。直到九年后，当我开始收集 1949 年前即开始行医的老医生们的口述史时，其中最意想不到的发现之一就是，这些医生在民国时期从事不少急性病的中医临床。例如，南京的周仲瑛回忆：在 1946、1947、1948 年天花流行期间，他曾经协助自己的父亲成功救治了许多病人[8]。他还记得他的父亲在 1946 年的霍乱流行期间常常使用五苓散通阳化气，救了许多病人的性命。[9] 南通的朱良春告诉我，他是如何在 1940 年的登革热流行中一举成名的。[10] 凭借着从两位老师那里学来的经验，他可以在三天内治愈自己所接诊的大部分登革热病人。[11] 生长于离上海不远处的崇明岛的沈凤阁，记得曾经陪自己的老师去最严重的病人家中诊治。“他们都患的是发热的急病，”他告诉我，“……像是肺炎、伤寒、痢疾、疟疾……老师的疗效很不错……他的威信很高。”[12]

起初，我对这些说法很是怀疑：这些急性病，当代的医生一定不敢只用中医治疗。我甚至想这些医生是不是把我当作了一个容易受骗的外国人，会轻易相信他们神乎其神的吹嘘。但是越来越多的老医生提到相似的病例，我开始反思自己原来对中医不擅长治急性病的偏见，再回过头去重新查阅这一时期的医学文献，这才发现大量之前没有注意到的类似病案。例如，不少孟河学派的名医以“和缓”的临床用药特点著称，主要迎合那些不愿用猛药快攻的娇贵病人，他们最初却是以治疗急性感染性疾病成名的。例如，巢少芳（1896–1950）因治疗脑膜炎和其他感染性疾病而出名；[13] 丁甘仁（1865–1926），民国早期最著名的医生和改革家之一，年轻时在上海艰难开业，1896 年终于靠成功治疗流行病烂喉痧而一举成名。[14] 随着访谈的深入，我愈发关注这一问题，与老医生们直接探讨民国时期中医治疗急性病的经验。哈尔滨的老医生张琪对此做了最好的总结：民国时期医生不总是在治疗急性病，但是“（民国时期）成名的中医都是以治疗急性病成名的”。

这些老医生的回忆所描绘的民国中医临床实践的图景，与我在半个世纪后的经历截然不同。随着采访的深入，我的研究问题开始转向，从中医可以治疗急性病这个令人吃惊的发现，转到中医如何在当代的急诊医学中逐渐变成一个无力的旁观者的现象。我的受访者并不能就这个问题给出简单的答案，他们自己也经历了这样的转变，而且每一位都在后来的职业生涯中因为治疗某种慢性病症而出名。这些受访者也都见证过西医学在急诊医学领域掀起的一些革命性发现——从对抗急性感染的抗生素，到在休克时维持血压的儿茶酚胺，到辅助呼吸的呼吸机，到能做出快速诊断评估的 CT 机。但是总的来说，这些访谈所揭示的是，中医的这种转变并不仅仅是由于西医在急诊医学的快速进步造成的，而在很大程度上是由于知识传承失败而引起的。

中医治疗急性病的知识（以及集体记忆）的缺失，在我一个临床带教老师所讲述的故事里得到生动的呈现。1999 年秋天，我在急诊科实习前不久，亦是我的口述史项目开始的十年前，我在大学附属医院的老年科实习。我注意到所有的住院病人有任何感染征象都要使用抗生素治疗。有一天，我问主治医胡医生，能否只用中药治疗这些感染。她不太确定，因为科里要求在有指征时使用抗生素。但她认为是有可能的，于是告诉了我一个她自己在院外治疗一个患细菌性肺炎的年长女士的故事。这位女士是熟人，不想住院，要求胡医生给她开一个处方。但当胡医生给她开出一个有抗生素的处方时，她拒绝了，要求中药处方。胡医生很吃惊。老年人得肺炎是死亡率很高的严重疾病，不能拿来开玩笑的。她睁大了眼睛讲这故事，“这个病人患大叶性肺炎还拒绝用抗生素，不是找死吗？但她很坚决，我没办法，只好给她开了一张中药处方。没想到，她居然好了！”胡医生被自己的成功震惊了。可惜当时胡医生和我都没有从这个故事中看到中医可以治疗“急病”的证据。

这个故事突显的集体记忆缺失，加上我那些受访者讲述的故事，揭示出当代中医在认识论上发生的深刻转变。为什么这么多医生、学者都没有意识到这些转变？托马斯·库恩、米歇尔·福柯和其他历史认识论学者的论述或许可以为理解这一问题提供启示。这些学者提出，认识论的快速转变不总是可辨识的，即便是对于那些参与促成转变的人来说也是如此。例如，库恩的作品反复揭示了许多科学探索领域的科学革命或者是他所谓的范式转换，都重塑了进步式的知识积累叙事——过去的错误最后引向了今日的正确理解。[15] 在本文中，我会主要基于受访者的说法来概括影响了这个认识论转变的社会和政治变化，同时，这种转变对于我、胡医生和无数的中医师来说是匪夷所思的：只用中医竟然也能治好大叶性肺炎。但是首先，我们必须重新审视民国时期医学的主流叙事者：迅速扩张的、现代化的西医学从业者和传统的、跟不上时代步伐的中医学从业者。只有我们认识到这种叙事的局限性，才可以探索促使医学领域转变的更有力的社会和人口力量。

三、论争的历史

民国时期医学史首先是中西医职业群体相互竞争的历史。[16] 这一时期冲突产生的主要原因是很多国人通过在国外的医学院校接受教育成了西医。这一时期也是帝国主义侵占中国领土的时期，这些医生通过自己的西方科学知识与中国社会的改革者结盟，尽管他们的数量很少，政治力量却相当不容小觑。一些医生与大多改革派领导一样对中医怀有敌意。在 1920 和 1930 年代数量激增的中医学杂志中，中医界对自己的新对手的政治潜能忧心忡忡。维护中医价值的争论性文章、关于政府官员歧视中医界的报道等等在这些杂志上时有出现。对于历史学家来说，1929 年，新成立的国民政府卫生部成员余云岫提出的废止中医法案使这一时期紧张的政治气氛达到巅峰。尽管这个法案没能通过，但它象征着中医界在 20 世纪大部分时间里所面对的偏见。

历史学者主要研究了民国时期这场冲突的两个特点。首先，拉尔夫·克罗泽尔和赵洪钧等学者研究了知识界关于中医价值的论争。受到“五四运动”激进主义的影响，许多中国知识分子在 1920 年代开始呼吁科学普及的急迫性。他们认为中医就是个他们实现中国社会现代化愿望的障碍，斥责其是旧社会的遗毒，是支撑着迷信的不科学实践[17]。如梁启超的著名嗟叹：“阴阳五行说为二千年来迷信之大本营……吾辈死生关系之医药，皆此种观念之产物。”[18] 最具影响力的批评来自余云岫，他在去日本学习西医之前做过中医。1916 年，他发表了一篇论文《灵素商兑》，该文批判了最古老和最受尊崇的医学著作《黄帝内经》，指责其“谬误可得而胜发乎？”而且是基于“粗率之解剖，渺茫之空论，虚无恍惚”[19]。

其次，雷祥麟和布赖迪·安德鲁斯等人指出了一些与这些知识界论争同时进行的政治斗争和制度发展。这两位学者的研究都表明西医学在中国的发展并非自然和必然。随着中国国民党在 1928 年组建了国民政府，这两个医学阵营开始激烈地争夺政府的资源和支持。西医医生们通过与政府的生命治理目标达成有效一致，获得了政治上的优势。[20] 而中医界也有效地组织起来阻止了 1929 年废止中医法案的颁布，通过结交国民党政府内部的同盟，在1930年代中期获得了一个薄弱却正式合法的与西医学平等的地位。[21]雷祥麟中肯地指出中医行业许多重要的改变都是由“与国家的相遇”（encounter with the state）激发的。他关于政府在这两个医学职业群体的冲突中所扮演的关键角色的论证，有助于解释为什么自1830年代西方传教士传播西医开始，西医学已在中国社会存在了几十年，却只对中医实践有很小的影响。

我的采访中一个最令人吃惊的发现就是，许多当时的中医师对于被描绘为“事关中医生死存亡”的废止中医大事件，或者漠不关心或是全不知晓。天津的李振华自1930年代十几岁时就开始跟随他的祖父在河北乡间行医。“我那时候很小，我甚至都不知道有两种医学，”直到多年后他的祖父去世，他到北京国医学院学医时才知道这一事实。[22]另一位同名的名医，来自河南洛宁县的李振华告诉我，他直到1949年中华人民共和国成立后才知道民国时期这段斗争的历史。[23]来自杭州的名医何任是家中的第三代医生，1938年到上海新中国医学院求学。“我们从来没考虑过为什么会学中医。我父亲就是干这个的，所以我长大也打算干这个……在那时候，中医医生和西医医生吵得不可开交，但我们并不知道这个事。”[24]

在我访问的39个医生中，只有两个医生似乎知道中西医论争的事情。生于1916年的邓铁涛，记得年少时在广州的报纸上跟进这场论争，当时读到有人说人参和白萝卜的临床效用一样时，很愤怒。他后来成为了广州中医专门学校的学生，听到亲自参加反对废止中医法案的老师所讲述的故事，对于这场争斗的意识和记忆才被加强。[25]邓铁涛在他漫长的职业生涯中一直热情维护中医的科学价值，所以他比别人更加熟悉这些辩论，我并不奇怪。但是我对干祖望的反馈有点吃惊，他生于1912年，是中医学耳鼻喉科专业化的关键人物。他向我讲述了他对余云岫的仰慕，还认为他的书《灵素商兑》是中医学最重要的书之一，而在中医圈内这书一般被斥为对中医学带偏见的、挑事儿的攻击。他很欣赏余云岫观点中的讽刺意味，他告诉我余云岫“侮辱了中医，但他是对的”。[26]

现如今这两个医学职业群体当年的争斗已众所周知，尤其是废止中医法案在中医学界激起的强烈政治反响，可是为什么这么多受访者对此含糊其辞，甚至根本不知晓？尽管1929年他们应该还太年轻，不可能参与那次重大的事件及其后的余波，但我以为他们都会像邓铁涛一样了解他们医学生涯开始时紧张的政治气氛。他们的漠不关心和一无所知提示我们，有必要对占据着民国医学史叙事主流的中西医论争进行重新评估：这些事件其实很局限，大概涉及少量大城市的精英分子，对全体中医从业者的冲击较小。这也正是干祖望会称赞余云岫的原因（他形成不了对中医的威胁）。余云岫是个富有洞察力的中医批判者，但是他从来没有能力，或许更重要的是国民政府也从来没有能力来执行他鼓吹的“医学革命”。历史学者赵洪钧详细讨论过这个论争，含蓄地认同了这个观点。赵洪钧指出这场两个医学系统的论争无疑是一个以上海为中心的城市现象。在北京和天津也有激烈的交锋，但这些城市的很多中、西医医生都支持两种医学体系的融会贯通。[27]我的受访者也许不是当时思想的最佳代言人——他们只是活下来的少数人中健康且愿意受访的那部分——但我想他们确实给出了与这个论争不同的叙事版本。[28]

如后文所述，至少就临床实践层面而言，这些受访者的其他回忆会继续取代以中西医论争为中心的叙事。他们关于民国晚期至共和国早期的人口统计、临床机构、医术传承特点的记忆，都提示了这一时期的另类历史解释，可以帮助我们理解为什么治疗急性病的中医实践，现在变成了专攻慢性病。

四、人数

民国时期大多数医生不关心中西医两个医学职业群体论争的一个关键因素，或许也是最主要的因素，就是中医师远比西医医生多。几乎我的所有受访者都说在他们周围，要么没有西医医生，要么很少。1931 年生于安徽歙县的李济仁说在他成长的过程中村子里一直没有西医医生。（那时候）在乡间，可以说 99% 的医生都是中医。[29]1924 年出生在北京城外密云县的郭中元回忆说，在他居住的小村子大兴庄有 4 个中医，而整个密云县只有几个西医。“他们在军队里当过一段时间的护士，然后回到家乡开了个诊所……他们的技术不怎么样……跟现在医院里的不同，他们也没有什么器械。只能拿个听诊器听一听……他们做不了什么，比起中医医生能治的病少多了。”[30]

在城里，西医医生多些，但还是明显少于相应的中医师的数量。西医聚集在医院里或是私人诊所中，他们的诊查和治疗通常超出大多数人的经济能力。[31] 政府资料估计，在 1937 年中日战争爆发时，全国大约有 9000 名注册西医医生，那时我的大多数受访者才刚刚开始学医或是从医。[32] 而大约 22% 的西医集中于上海，使得其他城市的西医医生数量更少。[33] 中西汇通派著名医家张锡纯，在他讨论 1929 年废止中医法案的一篇文章中，对两个医生职业群体的相对实力对比描绘了一幅类似的图景。

> 近闻中卫会议，诸位上峰，偏听西医之论说，欲废中医中药。此特因诸位上峰，非医界中人，不知中医与西医之实际也。即当时观之，都会商埠之处，病家延西医服西药者，不过十中之一。其在普通州县，病家延西医服西药者，实不过百中之二也。夫西医入中国已多年。使果远胜中医，何以信之者如此寥寥。此明征

也。且中医创自农轩，保我黄种。即遇有疠疫流行，亦皆有妙药灵方，为之挽救。是以我黄种之生齿，实甲于他种之人。[34]（标点符号为译者加）

即使在这篇文章发表20年以后，在新中国建立之际，张锡纯对于国家两种医学职业群体的概括依然相当准确。根据官方的数据，1949年西医医生数量增长了4倍达到了3.8万，而其中只有2万是从医学院毕业的。[35]那时相应的中医师的数据未可得，但1954年的内部资料估计有50万中医师，这个数据大致可以给我们一个民国末期医生数量的粗略印象。[36]

且不论确切的数字是多少，要点在于我的受访者们认为中医在民国时期医学实践中占优势。例如，何任记得杭州在民国时期有两所主要的西医医院，神爱医院和广济医院；这两所医院皆运行良好，但都无法与他父亲蒸蒸日上的诊所相比。他告诉我："大多数人更看重中医。"[37]邓铁涛注意到在抗生素发明以前，中医治疗急性感染性疾病的临床疗效甚至为西医医生所重视。他记得1940年代晚期救治过一个发烧的小男孩，男孩的父亲就在太平南路（今人民南路）上邓铁涛诊所的隔壁开了一家西医诊所。按照邓铁涛的说法，西医医生向他的中医竞争对手求助正是两种医学临床治疗能力对比的象征。[38]哈尔滨著名的针灸科医生张缙毕业于沈阳中国医科大学，1951年以西医身份开始自己的职业生涯。最初在临床工作时，他发现"西医只有极少治疗疾病的方法。那时候，有些人会开玩笑说我们是'三段大夫'。头是一段，如果头痛，我们就用阿司匹林。中间的胃肠道是第二段，如果胃疼，我们就用胃散（一种可能含有碳酸钙的化合物）……最后如果腿和手疼，我们就用保泰松。"[39]民国时期的马克思主义学者和中医师杨泽民曾试图就这种状况给出一个更加哲学化的解释："中医可以疗疾而不知病人之病；西医可能知道病灶所在而给不出疗法。这就是为什么中医病名混乱而西医少疗法。”[40]

五、私人诊所

第二个民国时期医疗实践的关键事实是私人诊所是主要的医疗服务所在地，西医院为数不多且集中在城市里。一些新成立的中医学院也建有医院作为临床训练基地，[41]但大多无法解决修建和运行医院的经费开支。[42]因此，即便那些上过中医学校的受访者——约占受访者总数三分之一——大多也是在他们老师的私人诊所接受临床训练而不是在中医医院里。一些城市医生另租地方开诊所，但大多数医生直接把诊所开在家里。农村地区的医生则还会在诊所里附带药房，因为买药不方便。城里的医生典型的执业方式是只看病开方不抓药，有时也做针灸治疗。通常上午门诊，下午上门出诊到那些或身体虚弱无法自行就诊、或足够富裕能够支付额外费用的病人家中。

由于医疗实践是私人性质而且病人是自费看病，经济因素对临床工作的影响很大。我的很多受访者都说一般病人只在得急症时看医生。这就意味着许多迁延、慢性、需要长期治疗的病情，或是不太严重、能慢慢自愈的小病，皆得不到治疗。这些病只有富人才看得起。普通人只在急性病时才来就诊，期待几剂药就能显效。著名药师金世元1940年代早期在北京的复有药庄做学徒，他回忆说：“（不像现在）没听说过一个方子上开十付药的。我们抓的最大的方子也就是两到三付药，经常只有一付药。”[43]娄多峰十几岁的时候在河南的原阳县跟随他的祖父学医，后者善治温病。“他从来都在三付药内治好一个病人，经常一到两付药就好了。”[44]李今庸生长在湖北山区的枣阳县，他指出战争带来的贫困（因为日本人和国民党军队的抢掠）使得农民根本无法顾及急病以外的病症，同时贫困又为流行病的滋生创造了条件。“那时候抗战时期，乡下的生活太困难了。农民只吃得起一两付药。我们只治急性病。你一两付药不见效，病人不会再来的。”[45] 只有少数几个受访者，像来自广东惠州的黎炳南和来自辽宁丹东的周信有，说他们看了相当数量的慢性病人。[46] 究其原因，前者是因为南方较为富有，而后者则是由于日本人的殖民地政策，急性、感染性的疾病都要送到西医医院去救治。[47]

六、医学训练

关于民国时期医疗实践的第三个关键事实是医学训练。尽管已经有了现代学校建制的教育体系的迅速传播，这一时期的医学训练还是以学徒制和医学经典为核心。我的受访者中大约三分之二是通过师从亲戚或本地老师学医的。余下的三分之一上过民国时期的新事物——中医学校。[48] 但这些人在完成他们的学业后几乎都做了某个临床老师的学徒或类似学徒的工作。阎润茗于北京的华北中医学院毕业后又跟随赵树屏做了五年学徒，同时也跟随一位佛教僧人李春仙学习针灸。[49] 何任在上海新中国医学院的最后一年准备进入临床训练时，选择了回家跟随父亲实习。

学徒和学生使用的课本有些区别，但与同时代的学生课程之间的区别比起来，这些区别还是很小。学徒的主体教材基本没有偏离 19 世纪的标准，一是类似启蒙读物的组合，比如《药性赋》、《本草备要》、《汤头歌》，然后就是经典，受过教育的老师尤其强调后者。北方的受访者大多读清代的经典《医宗金鉴》。而在南方则要求学徒修习“四部经典”，对温病学派的方法相当注重。对于学徒训练来说，死记硬背这些书籍是很重要的，这些受访者今天依然能背诵出 70 多年前记下的段落。而在新式的中医学院，学生的教科书是他们的老师编写的。尽管教科书的复杂编写过程超出了本文的讨论范围，总的来说，它们是对通行的入门书籍和经典的简单编辑，有时带有非常个人化的解读。相较于新中国早期集体编写、今日中医院校课程基础的标准教科书而言，这些民国时期的教科书彼此之间有更多共同之处。[50]

也许更重要的是，这些受访者当时只接受过很少、或完全没有西医训练。这与当今的中医大夫形成鲜明的对比。今日的中医师不仅要大量学习西医，还必须能够胜任西医临床实践。在民国时期，西医知识只受到一小部分医生如张锡纯、恽铁樵、陆渊雷等人的认可，他们希望改良中医使之“科学化”。尽管这些医生的写作受众甚广，对年轻医生的影响也很大，这一理念只限于精英圈子，未被传播到教育改革中。（民国时）新近成立的中医学院都有西医学的课程，但是通常不会超过解剖和生理这样的基础课。在这些私立学校求学的受访者们也表示这些课程非常基础。其他以学徒形式学习中医的受访者，基本上没有接触过西医学，或许偶尔读到晚清医生唐宗海的著述中包含的西医解剖知识。这些回忆表明，民国时期的普通医生不需要知道西医知识，这与今日的中医师形成鲜明的对比。

还有一些例外来反证这个常理。抗日战争爆发后，朱良春到上海跟师中医名家章次公完成他的医学训练。他回忆章次公会让病人去到一个临近的实验室做血液生化检查，提倡“双重诊断，一重治疗”，也就是做出中医和西医的两个诊断，但只用中医进行治疗。朱良春追溯自己之所以致力于如今叫作“中西医结合”的专业和这段经历紧密相关。跟随一个走在医学新流派前沿的老师学习，他依然记得当时的欣喜。但他也记得当时大多数医生认为他的老师是中医的叛徒，章次公的回应是：“并不是我想要做中医的叛徒”，“时势造叛徒。”[51] 饱受赞誉的第五版《方剂学》主编之一许济群也怀旧地忆起早年间在上海学了一些西医。他上了一个关于“三常规”的特殊课程，于是在 1940 年代就把血、尿和痰的化验结合到自己的执业实践中。他骄傲地告诉我，他曾经使用这些技巧诊断了一个恶性疟疾的病例，用奎宁治好了病人。我问他使用西医方法时是不是感觉“背叛”了中医。他摆摆手否认了那种说法。他回忆说那时候上海的医疗市场竞争极其激烈，所以先人一步很重要。[52] 临床结合西医在新中国是中医实践的常规，在民国时则相当稀罕。

七、与国家政权相遇

上文总结的民国时期中医学实践的三个方面——中医医师的数量优势，医院数量的稀少（中医医院近乎没有），以及与西医学最低限度的接触——在新中国随着政府开始全面建构国家医疗服务体系而迅速改变。随着社会环境的改变，中医临床工作的性质也发生了改变。在 1950 年代，中医师确曾在控制一些流行病时起到过重要作用，最知名的就是 1955 年在石家庄和 1956 年在北京的日本乙型脑炎爆发。[53] 第二次爆发使得蒲辅周跻身成为同时代的中医大家之一。当时蒲辅周认识到石家庄疫情的治疗策略不如北京控疫的效果好，迅速起草了一个简要的论述，提出 8 种不同方案和 66 种治疗日本乙型脑炎的配方，不仅改善了临床预后，还完美结合了民国时期针锋相对的伤寒和温病两个学派的方法。[54] 但是，尽管取得了这些成就，对中医的偏见也在逐渐成形，路志正回忆说，其后不久便经常听到“中医不能治感染性疾病”[55] 的言论。根据邓铁涛的说法，在 1950 年代晚期，中医师已经开始把注意力从急性病转向了慢性病。[56]

这一走向的令人惊讶之处在于，彼时正是政府对中医的投入达到空前高度之时。新政府经常把自己的中医政策描述为与国民党的压制政策相反，尤其以成立了30所中医院校、建立了2500所中医院，训练了成千上万的中医师为傲。[57] 这些成就与民国时期相比确实了不起，毕竟那时是个人而不是政府在尽力发展中医的现代体制。但是自从1950年代以来，中医师的地位下降了，他们的临床工作范围被蚕食了。我们该如何解释这些发展呢？借用雷祥麟的句子，我会说1950年代才是中医与国家相遇的时期，而不是雷祥麟所说的民国时期；中医师们不是作为政治行动者争取国家资源，而是积极参与了国家医疗服务体系基础设施的建设。周信有关于他在伪满洲国（一个极力推崇西医的政权）和新中国辽宁省从医生涯之不同的说法应该可以适用于全中国："（解放前）没人干涉西医和中医医生之间的事。尽管政府不支持中医，但它还是允许你自己开业。然后（解放后）干涉就来了，西医"领导"中医了。"[58]

这种"领导"有多种形式。首先，建国初期投入了大量资金用于建设西医机构。根据卫生部的官方数据，1949年建国后西医医生的数量快速增长。从1949年统计的3. 8万到1957年的7. 36万人，8年间西医人数几乎翻了一倍。在接下来的8年更是以两倍半的速度增长，到1965年达到了18.87万人。[59] 如此快速增长所付出的代价是医生质量的下降，有观察者指出医学院校的班级曾经有过每班400–600名学生的情况。[60] 相比起来，1957年官方统计的中医医师有33. 7万人。从当年估算的50万人骤减至此有可能是新的执照制度所致。[61]1957年，药师金世元也参加了其中的一次执业考试。据他回忆，当年参加考试的1900多人只有150人获得通过。他将如此高的失败率看作或多或少真实地反映了应试者的技术水平，而不是因为（有关部门）企图缩减中医从业人员的规模。可是在1950年代早期，有学问的门外汉尚可通过学习中医谋生，自学医书然后自命为医生。[62] 无论执业考试背后的政治动机是什么，结果是中医界业内还在筛选良莠的时候，西医的职业发展已经遥遥领先了。

与此同时，中医的机构建设也更为谨慎，尤其在1950年代早期。直到1954年，针对中医的政策都是强调“中医科学化”，意思是对现有中医的重新培训而不是发展新的机构。自1950年始，直到1950年代后期，政府极力鼓励中医师参加“中医进修班”，主要学习西医学基础。[63]进修班通常下午或者晚上上课以配合医生们的门诊安排，为时6个月到1年不等。除了西医培训，政府还鼓励医生们学习马克思主义和毛泽东思想，有时候还要去上正规的学习班。尽管我的大部分受访者抱怨这一时期官僚对中医的偏见，但他们大多对于有机会学习西医和马克思主义持肯定态度。河南的李振华自学了辩证唯物主义，他告诉我，“这是理解内经的关键。”[64]在这一时期，中医师并不在医院工作。政府鼓励他们组建“联合诊所”，通常由不到十个医生结成小组集体开业。我的大部分受访者认为这一方式对职业有利，因为它既促进了相互交流，还能减轻独自开业的经济压力。

从1954年开始，卫生部的一些高层干部被免职后，政府改变了政策，中医的机构建设急切地开展起来。[65]但是对中医科学化的强调仍然以新的形式继续着。一个新的做法是改为“西医学习中医”，由此训练两种医学都擅长的专家，希望他们能够找到两种医学的结合点而实现“中西医结合”。为期3年的“西学中”班自1955年开课，中西医结合课程的一些形式一直延续到了今天（尽管近年来势头渐衰）。当年一些西医大夫是不愿意参加这些实验性课程的，但也有许多沿着这条路走下去而成为了中医界的引领者。第二种改造中医的方式是组建新的中医学院。民国时期成立的私立中医学校，只有一所在抗日战争的经济危机和战后国民党政府对中医的压制中幸存了下来。1956年，中央政府在北京、上海、广州和成都成立了四所中医学院，很快扩展到其它主要的省会城市，到1965年，共成立了21所中医院校。学习西医成为这些新的中医学院课程的中心内容，占了将近50%（医学相关）的课时数。尽管大多数中医教育家认可接受一定西医学训练的必要性，但中医与西医课程的恰当比例是争议的重点，最后引发了当时几所新成立的中医学院的5位教授带头给卫生部上书的著名事件——“五老上书”。五老抱怨中医与西医一比一的课程占比会造成极大的中医教学问题：想要掌握两种医学，最终一个也不精通。[66]

与成立这些中医学院同步的是中医医院的建立。基于1950年代早期“联合诊所”的经验并利用西医医院的经营模式，这些中医院在标准的医院工作中结合了大量的西医学技术，和民国时期的中医院很是不同。像邓铁涛回忆的，在那些日子里，所有的病人都得有个西医学的诊断。也有一些西医医生被分配到这些医院来完成这一任务。[67]在著名的中医骨科专家诸方受看来，西医学在新的中医院的存在感也在其他方面持续增强。诸方受一度参与了北京医学院（1952-1957）的一个实验项目，对中医师进行西医综合训练。1957年，这些学生完成学业时，将额外的西医学技巧带到了这些新的中医院。接下来的一年，“西医学习中医班”的第一届毕业生进入了医院，接下来是下一年的第二届毕业生。在1962年，新的中医学院的第一届毕业生带着他们学到的大量西医学知识开始在这些医院工作。[68]

新中国中医学院及医院的创建对于中医学界来说是很重要的成就，这些成就大概是民国时代医生根本无法想象的。但是这些制度上的成就同样把西医带入了日常中医临床实践，并最终导致了两种医学体系的结合，这种结合又成了今日中医的特点。[69]方小平指出，类似的西医送医下乡运动导致了灾难性的后果，极大地破坏了中医在农村的实践。建国初期，政府相当倚重中医师在乡间组成联合诊所。但是联合诊所很难招到新的医生，部分原因是学徒制的传统学医方式在新时期的社会条件下难以为继。1949年前，富裕、有条件的农家可能会鼓励孩子去跟师学医；到了1950年代，这些家庭是阶级斗争的典型目标，无法再资助他们的孩子学医。[70] 由于新的中医师有限，联合诊所逐渐引入了一些接受过西医训练的新人。据方小平的调查，到1960年代早期，杭州地区联合诊所的中医师比例已经不足50%。[71] 随着“文化大革命”（1966–1976）的到来，赤脚医生项目的启动加速了这一进程。该项目的本意是立足当地、解决农村地区医疗服务的普遍不足。但是与政治宣传相反，方小平指出杭州地区的大多数赤脚医生是在外地接受培训，而且主要是西医训练，这些医生也首先把自己看作西医的从业人员。[72] 到“文革”晚期，这些新的培训流程和农村药物普及性的逐步提高使得西医在中国农村的医疗实践中占据了主导地位。

经过三十载名义上支持中医、但实际上限制中医并在体制上强行推进中西医结合模式的政策的侵蚀，“文化大革命”终结之时，中医业可能也跌到了谷底。历史还不能完全告诉我们“与国家的相遇”对中医来说意味着什么。但或许身为中医师同时也是政府高官的吕炳奎，其毕生的政治生涯都在积极推动中医业的发展，可以帮助我们理解毛泽东时代社会和政治改变对中医的影响。粉碎“四人帮”后不久，允许对“文化大革命”进行批判时，吕炳奎描绘了中医领域的一个空白画面，称之为“十年浩劫”。

> 十年浩劫之后只有24万，现在25万（中医大夫），同解放初期相比去掉了一半。西医发展了十多倍，中医去掉了一半。这个“一半”，在统计数字上看是一半，实际不止。我们调查的材料，20几万中医只有20%–30%系统学过中医……中医人数这样少，中医机构也少的可怜，全国将近200万张病床，中医只有5万张，这5万张床位，真正搞中医的床位大概还不到5千张……现在中医只能看一点门诊，一点普通的常见病，这能总结东西吗？能提高吗？是不可能的。[73]

尽管有吕炳奎的悲观论调，1980年代开始中医有了显著的复兴。而在2000年左右，我的许多受访者也感觉中医的前景达到了他们记忆中前所未有的光明程度。但是，中医学界自1980年代起将自己重建成为与过去相当不同的医学实践形式。其中改变最明显的方面就是它成为了“慢医学”。

究竟为什么毛泽东时代的社会和政治改变使得中医大夫们转向治疗慢性病？根据邓铁涛的说法，总体原因是中医师治疗急性病的“舞台被夺走了”。他指出1950年代初西医医院数量的急速增长以及1951年实施的只提供住院费用报销的医疗保险制度的开创，将享受医保的病人都拉进了西医医院。西医院的技术——实验室检查、X线检查和其他的医学器械——令人印象深刻，这些患者便逐渐倾向于选择西医治疗急性病。[74]李金庸指出在武汉，医院管理者们在推进这种转变中起到了核心的作用。因为医院管理者通常是西医背景，他们更能敏锐察觉（甚或支持）他们在卫生部的上级对西医的偏好，而他们所处的位置又可以依照自己的偏见缩减中医师的工作。

早在1950年代日本乙型脑炎爆发期间，中医师们在表现出有能力对这样一个公共卫生危机状况做出贡献的同时，也发出了怨言。在一本河北卫生工作者协会1956年出版的手册的引言中，协会的主席同时也是河北卫生局局长段慧轩赞扬了中医师对日本乙型脑炎治疗做出的贡献，并抱怨西医医院阻碍了他们的工作：

> 一方面是宣传中医的治疗经验，但是更重要的一方面是与（西医）医院的紧密合作，才能使得中医经验更好的得到施行……一些医院不太配合，影响了中医治疗的使用。一些医院在允许中医治疗之前过度强调了西医的初步诊断，引起了（病人治疗的）不必要的延误，这是很不恰当的。现在证明了中医学治疗脑炎有

> 很好的效果，我希望所有的机构都能按照革命人道主义精神，把病人的生命放在首位，通力合作，尽量有效地帮助中医师们。[75]

可惜当时重西医、轻中医的社会和政治力量压倒了中医治疗急性病的乐观情绪。像李今庸说的，在日本乙型脑炎爆发中所展示的如此受瞩目的临床精湛技艺没能传递到下一代医生的手中。

> 为什么如今（年轻医生的）临床技艺比不上老一辈的？老一辈医生积累了几十年的经验，自旧社会以来他们见了很多病例。但是……自从第一届毕业生到医院工作开始，无论何时有患急性热病的病人，病人马上就被送去西医治疗。中医师不允许参与……结果，老医生没法在治疗急性病中使用他们的技巧，年轻医生们也无法学到它们。[76]

八、尾声

2003年初，SARS的流行横扫中国和全球许多地区。中国的医疗机构在奋力救治这一未知的、高传染性、高致命性疾病的病人时，中医治疗急性病的作用在危急关头显现了出来。韩嵩注意到了在控制这场流行病的过程中，中医师扮演的积极角色几乎完全被西方媒体所忽略。[77]在危机开始之时，政府远没有认识到中医可以在控制这场流行病中起任何作用。李今庸告诉我，湖北中医药大学附属医院的领导们因为担心医院无法治疗SARS而不敢接纳一个SARS的病人。[78]在主要的疾病暴发区北京，官方指定了六个救治中心，全部都是西医医院，还关闭了中医的东直门医院，因为它是这个城市最早的几个病例出现的地点。[79]

中医得以进入 SARS 治疗可能由于流行病的爆发地恰好是广东省。今天中医在这个省份比在中国其他任何地区都更受欢迎。在省会广州，中医师从一开始就加入了治疗 SARS 的队伍。邓铁涛，全城最受尊重的名中医之一，骄傲地告诉我正是因为中医的贡献，广州的 SARS 死亡率比全国其他地方都低。在一项对 2003 年 1 月到 4 月的 103 例收入广东省中医院的 SARS 病人的回顾研究中，研究者发现死亡 7 例，死亡率为 6.79%，低于其他流行区域高达 15% 的死亡率。[80] 邓铁涛坚持认为尽管这些统计数据已经很引人注目，仍然不能说明全部的事实，因为他们忽略了所有尚未发展到可以确诊就被中草药及时治愈的高热病人。这些数据也没能显示预防性服用中药的医院工作人员中没有一例 SARS 感染，这也提示了中医的另一个优势——重视预防。[81]

广东省中医院的治疗是中西医结合治疗的疗法。而西医的主要治疗（除了标准的急诊和重症监护措施外）采用大剂量的类固醇，这类药物不但作用有限，有时还会导致毁灭性的副作用。在没有合适的西医治疗方法的情况下，医生们重新发现了中医已有的诊断和治疗急性病的精细理论体系。中医师在广州的成功最终在其他爆发地区也得到了认可。当年 5 月 3 日，香港官方邀请广东的中医师帮助他们控制疾病流行。广州中医师的积极作用最终引起了北京中央政府的注意，数个研究机构动员起来进一步研究他们的治疗成果。当年 5 月 8 日，当温家宝总理宣布“在防治非典中要充分发挥中医的作用”后，中医师全员加入了对抗 SARS 的战斗。[82] 对邓铁涛来说，作为那个已经被遗忘的时代的健在代表之一，SARS 危机唤醒了那个久远的中医师靠着他们治疗急性病的能力成名的时代。他对我说，中医大夫“终于又一次知道了他们是可以治疗急性病的”。[83] 不知年轻一代的医生们是否可以不辜负他的期望。

（赖立里　谷晓阳　译）

致谢

本研究受到美国学术团体协会及其美国研究中国人文学学者计划的支持，特此致谢。

注释

[1] 中国卫生部,《2011 中国卫生统计年鉴》, http: //www. moh. gov. cn/htmlfiles/zwgkzt/ptjnj/year2011/index2011. html. Accessed June 27, 2014。

[2] 广州中医药大学，广州中医药大学第一附属医院 http: //www. gztcm. com. cn/Default. aspx?tabid=99. Accessed July 2, 2014。

[3] Eric I. Karchmer, 2010 "Chinese Medicine in Action: On the Postcoloniality of Medicine in China", *Medical Anthropology* 29 (3): 226–252.

[4]《中国卫生年鉴》编委会,《中国卫生年鉴 2001》，北京：人民卫生出版社，2001。《中国卫生年鉴》编辑委员会,《中国卫生年鉴 2002》，北京：人民卫生出版社，2002，第 454–455 页，499 页。

[5] 黄星垣本是西医出身，后作为"西学中"人员参加了中西医结合班的培训。他的双重医学背景使他成为推动中医急症研究的理想人选。见黄星垣等主编，《中医内科急症证治》，北京：人民卫生出版社，1985 年。

[6] 当时有几位西医大夫在医院工作。他们看似无声的存在背后有建国初期成立中医院的历史，西医大夫是被调来帮助中医同行们熟悉建制化的医疗体系的。

[7] 国家中医药管理局医政司,《全国中医医院急诊必备中成药应用指南》，国家中医药管理局医政司，1997，第 1–2 页。

[8] 与周仲瑛的个人访谈，南京，2009 年 1 月 16 日。

[9] 王志英等，"走近中医大家周仲瑛",《医学人生丛书》，北京：中国中医药出版社，2008 年，第 16–18 页。

[10] 与朱良春的个人访谈，南通，2008 年 12 月 22 日。

[11] 曹东义等，"走近中医大家朱良春",《医学人生丛书》，北京：中国中医药出版社，2008 年，第 92–97 页。

[12] 与沈凤阁的个人访谈，南京，2009 年 3 月 15 日。

[13] Volker Scheid, *Currents of Tradition in Chinese Medicine 1626–2006* (Seattle: Eastland Press, 2007), 150.

[14] Ibid., 228.

[15] Thomas S. Kuhn, *The Structure of Scientific Revolutions*, 2nd ed. (Chicago: University of Chicago Press, 1970); Michel Foucault, *The Order of Things: An Archaeology of the Human Sciences*, 1st American ed. (New York: Pantheon Books, 1971); Arnold Davidson, *The Emergence of Sexuality: Historical Epistemology and the Formation of Concepts* (Cambridge & London: Harvard

University Press, 2001).

[16] 邓铁涛等主编,《中医近代史》, 广州: 广东高等教育出版社, 1999; Ralph C. Croizier, *Traditional Medicine in Modern China: Science, Nationalism, and the Tensions of Cultural Change*, Harvard East Asian Series, 34 (Cambridge,: Harvard University Press, 1968); 赵洪钧,《近代中西医论争史》, 石家庄: 中西医结合研究会河北分会, 1982; Sean Hsiang-Lin Lei, "When Chinese Medicine Encountered the State: 1910-1949" (Ph. D. dissertation, University of Chicago, 1999); Bridie Jane Andrews, "The Making of Modern Chinese Medicine, 1895-1937" (Ph. D. dissertation, University of Cambridge, 1996).

[17] Croizier, Traditional Medicine in Modern China: Science, Nationalism, and the Tensions of Cultural Change: 72-77.

[18] 赵洪钧,《近代中西医论争史》, 第225页。

[19] 余云岫,《医学革命论集》, 上海: 大东书局, 1932 (1928), 第1页。

[20] Lei, "When Chinese Medicine Encountered the State: 1910-1949"; Andrews, "The Making of Modern Chinese Medicine, 1895-1937".

[21] 邓铁涛等,《中医近代史》, 第177页。

[22] 与李振华的个人访谈, 天津, 2008年12月15日。

[23] 与李振华的个人访谈, 郑州, 2009年3月30日。

[24] 与何任的个人访谈, 杭州, 2009年4月2日。

[25] 与邓铁涛的个人访谈, 广州, 2009年3月19日、2011年6月16日。

[26] 与干祖望的个人访谈, 南京, 2009年1月17日。

[27] 赵洪钧,《近代中西医论争史》: 第98-101页。

[28] 我走访了中国的大部分地区, 包括东北、华北、华中、长三角地区和四川、广东省, 只有为数不多的医生经历过这一时代。其中有一些医生在大城市长大, 大多是在农村开始行医的。

[29] 与李济仁的个人访谈, 北京, 2009年4月。

[30] 与郭中元的个人访谈, 保定, 2008年12月28日。

[31] 邓铁涛等,《中医近代史》, 第15页; Croizier, *Traditional Medicine in Modern China: Science, Nationalism, and the Tensions of Cultural Change*: 52.

[32] Croizier, *Traditional Medicine in Modern China: Science, Nationalism, and the Tensions of Cultural Change*: 54-55.

[33] Scheid, *Currents of Tradition in Chinese Medicine 1626-2006*: 182.

[34] 张锡纯, "中西医治疗上之真实的比较",《医界春秋》35(1929)。

[35] 崔月犁等,《新中国中医事业奠基人: 吕炳奎从医六十年文集》, 北京: 华夏出版社, 1993年。

[36]《中医工作文件汇编》编辑部,《中医工作文件汇编(1949-1983年)》, 北京: 卫生部中医司, 1985年。

[37] 与何任的个人访谈, 杭州, 2009年4月2日。

[38] 与邓铁涛的个人访谈, 广州, 2009年3月19日。

[39] 与张缙的个人访谈，哈尔滨，2009 年 3 月 5 日。
[40] 杨则民主编，《潜庐医话》，北京：人民卫生出版社；赵洪钧，《近代中西医论争史》，第 239–240 页。
[41] Scheid, *Currents of Tradition in Chinese Medicine 1626–2006*: 196.
[42] 邓铁涛等，《中医近代史》第 175 页。
[43] 与金世元的个人访谈，北京，2009 年 4 月 3 日。
[44] 与娄多峰的个人访谈，郑州，2009 年 3 月 31 日。
[45] 与李今庸的个人访谈，武汉，2009 年 4 月 1 日。
[46] 与黎炳南的个人访谈，广州，2009 年 3 月 19 日；与周信有的个人访谈，兰州，2009 年 3 月 29 日。
[47] See Michael Shiyung Liu, *Prescribing Colonization: The Role of Medical Practice and Policy in Japan-ruled Taiwan*. Ann Arbor: Association for Asian Studies, 2009.
[48] 这个比例不能代表当时医生的整体情况，毕竟那时受过学院式教育的中医大夫少之又少。
[49] 与阎润茗的个人访谈，2008 年 12 月 16 日。
[50] Eric Karchmer, "Orientalizing the Body: Postcolonial Transformations in Chinese Medicine" (Dissertation, University of North Carolina, 2005).
[51] 与朱良春的个人访谈，南通，2008 年 12 月 22 日。
[52] 与许济群的个人访谈，南京，2009 年 1 月 18 日。
[53] 中国中医研究院，《中医药防治非典型肺炎（SARS）研究（二）》，北京：中医古籍出版社，2003 年，第 43、48、59 页。
[54] 见蒲辅周、高辉远，《中医对几种急性传染病的辨证论治》，北京：人民卫生出版社，1960 年，第 51–64 页；Eric I. Karchmer, 2013 "The Excitations and Suppressions of the Times: Locating the Emotions in the Liver in Modern Chinese Medicine," *Culture, Medicine, and Psychiatry* 37 (1): 8–29.
[55] 中国中医研究院，《中医药防治非典型肺炎（SARS）研究（二）》，第 43 页。
[56] 与邓铁涛的个人访谈，广州，2009 年 3 月 19 日。
[57]《中国卫生年鉴》编辑委员会，《中国卫生年鉴 2001》，第 454–455 页。
[58] 与周信有的个人访谈，兰州，2009 年 3 月 29 日。
[59]《中国卫生年鉴》编辑委员会，《中国卫生年鉴 2001》，第 455 页。
[60] Victor W. Sidel, "Medical Personnel and Their Training", in *Medicine and Public Health in the People's Republic of China*, ed. Joseph R. Quinn (Washington, D. C.: National Institutes of Health, 1973), 156.
[61] Kim Taylor, *Chinese Medicine in Early Communist China, 1945–1963: A Medicine of Revolution*, Needham Research Institute Studies Series (London: Routledge Curzon, 2004). 37–41.
[62] 与金世元的个人访谈，北京，2009 年 4 月 3 日。
[63] Taylor, Chinese Medicine in Early Communist China, 1945–1963: A Medicine

of Revolution: 38–41; ibid.

[64] 与李振华的个人访谈，郑州，2009 年 3 月 30 日。

[65] Taylor, *Chinese Medicine in Early Communist China, 1945–1963: A Medicine of Revolution*; David M. Lampton, *The Politics of Medicine in China: The Policy Process, 1949–1977*, Westview Special Studies on China and East Asia (Boulder: Westview Press, 1977).

[66] 任应秋，《任应秋论医集》，北京：人民卫生出版社，1984 年，第 3 页。

[67] 与邓铁涛的个人访谈，广州，2009 年 3 月 19 日。

[68] 与诸方受的个人访谈，南京，2008 年 12 月 21 日。

[69] Volker Scheid, *Chinese Medicine in Contemporary China: Plurality and Synthesis*, ed. Barbara Herrnstein Smith and E. Roy Weintraub, Science and Cultural Theory (Durham, NC: Duke University Press, 2002); Karchmer, "Chinese Medicine in Action: on the Postcoloniality of Medicine in China" .

[70] 联合诊所在中医知识传承中究竟扮演了怎样的角色，并不清楚。方小平也谈到，在杭州郊区的姜村，那些有可能学医的村民已经不愿忍受跟师学医的辛苦。见 Xiaoping Fang, *Barefoot Doctors and Western Medicine in China* (Rochester: University of Rochester Press, 2012)，第 44、49 页。另外，我的受访者也提到类似的印象。比如李今庸曾经提到当年他父亲在湖北农村的好几个徒弟没能完成学医而中途放弃（与李今庸的个人访谈，武汉，2009 年 4 月 1 日）。

[71] 同上书，第 63 页。

[72] 同上书，第 53–66 页。

[73] 崔月犁等，《新中国中医事业奠基人：吕炳奎从医六十年文集》，北京：华夏出版社，1993 年，第 67 页。

[74] 与邓铁涛的个人访谈，广州，2009 年 3 月 19 日。

[75] 河北省卫生工作者协会，《流行性乙型脑炎》，保定：河北人民卫生出版社，1956 年，第 3 页。

[76] 与李今庸的个人访谈，武汉，2009 年 4 月 1 日。

[77] Marta Hanson, "Conceptual Blind Spots, Media Blindfolds: The Case of SARS and Traditional Chinese Medicine" , in *Health and Hygiene in Chinese East Asia: Publics and Policies in the Long Twentieth Century*, ed. Angela Ki-Che Leung and Charlotte Furth (Durham: Duke University Press, 2010).

[78] 与李今庸的个人访谈，武汉，2009 年 4 月 1 日。

[79] 邓铁涛等主编，《中国防疫史》，南宁：广西科技出版社，2006 年，第 702 页。

[80] 林琳，杨志敏，邓铁涛，"中医药治疗 SARS 的临床研究"，载《邓铁涛学术思想研究（Ⅱ）》，徐志伟、彭炜、张孝娟主编，北京：华夏出版社，2004 年。

[81] 与邓铁涛的个人访谈，广州，2005 年 8 月 13 日。

[82] 邓铁涛等主编，《中国防疫史》，第 703 页。

[83] 与邓铁涛的个人访谈，广州，2009 年 3 月 19 日。

Surra and the Emergence of Tropical Veterinary Medicine in Colonial India

James L. Hevia*

Surra is a name long used by Punjabi *sarwans* (camel-handlers) for a progressive wasting disease that kills or debilitates camels. It is also deadly to equines and a variety of other domestic animals. Since the late nineteenth century, veterinary scientists have considered the agent of the disease to be *Trypanosoma evansi*, a flagellated protozoan blood parasite dependent on a biting insect vector for its spread. *T. evansi* is one of a number of species that make up the genus *Trypanosoma*, the members of which are distributed widely across the tropical regions of the world. Species can be found in numerous African, Asian and South American domestic and wild animals, and at least two types have become pathogenic in human populations in the form of African Sleeping Sickness (*T. brucei*) and Chagas disease (*T. cruzi*), which occurs in Latin America. In all of their hosts, trypanosomes cause anemia, edema,

* James L. Hevia, Department of History and the New Collegiate Division, University of Chicago.

and emaciation, and the disease is nearly always fatal. In British India, surra only came into focus as an animal health issue after 1880, when veterinary researchers demonstrated that a flagellated protozoa was a primary cause of death among Indian Army cavalry horses and pack animals such as camels, ponies, and mules.

This paper considers how surra became an object of strategic military and economic concern for the colonial state, and how the interventionist policies of the state sought to control the disease while simultaneously, if inadvertently enabling its spread in Northwest India. In order to explain why this was the case, it will be necessary to deal with a set of internal developments in India influenced by colonial science in other parts of Asia and in Africa. In India, the spread of surra and the resulting death of hundreds of camels, horses and mules can be accounted for in part by the incoherence of colonial policy. From the 1890s forward, units such as the Army Supply and Transport Department, the Punjab Public Works Department and the Punjab Forestry Department often found their "improvement" initiatives in the northwestern India working at cross purposes, with the policies of one unit undermining those of others. As a result, competing policies made it difficult for veterinary surgeons to limit the impact of surra.

Outside of India, major changes in disease theories and scientific investigative methods in Europe and tropical Africa influenced research methods in human and animal medicine in India. These developments also resulted in an almost simultaneous emergence of the field of tropical veterinary medicine across Africa and Asia. British veterinarians in India found themselves enmeshed in an actor network of scientific expertise that they both learned from and made contributions to. Timothy Mitchell was right to point to a "rule of experts" in British colonies from roughly the 1890s forward. [1] But one might argue that this was a rule throughout the colonized world. More importantly, different kinds of experts — hydraulic engineers, human and agricultural statisticians, animal and land managers, and medical scientists — did not always sit comfortably side-by-side; their proximity produced conflicts, competitions, confusion, and contradictory policies and practices.

The Improvement State and the Transformation of the Punjab[2]

In the period between 1892 and 1914, the ecology of the Punjab was radically altered by the interventionist projects of hydraulic engineers, veterinary scientists, and professional civil and military administrators. As a result, relations between animals, plants, insects, micro-organisms, and humans were fundamentally and permanently changed. Economically, the colonial state was seeking to expand the region's irrigation works in order to eliminate "wasteland" , curtail pastoralism, and increase agricultural production in the Punjab, particularly of cash crops. [3] Militarily, the state was attempting to create a stronger presence on the Afghan frontier by, among other things, addressing long-standing logistical problems of frontier warfare through the creation of a permanent army transport system based on animal labor. The transport system developed at this time was part of a larger assemblage made up of rail and telegraph communication lines, mobile cavalry units, and frontier cantonments. All of these formations together composed a professional military security regime on the Northwest Frontier of India. [4]

Lastly, in order to insure a fit supply of pack animals for army transport, the state created mechanisms for intervening into the life of certain animal species such as camels and mules. Beginning in 1892, these interventions were carried out by the newly created Civil Veterinary Department (CVD), whose duties also included overseeing the breeding of horses and mules in India for army service. The new department was, in turn, made up of veterinary surgeons whose younger members were under the influence of the germ theory and laboratory methods then being developed in France and Germany, seeing them as technologies of knowledge for understanding and combating epidemic diseases in the working animals in their charge. This emphasis on what came to be called microbiology, with its subfields of bacteriology and parasitology, would alter the way veterinarians looked at animals and how they interacted with them. Put simply, animals became populations to be managed, while many sick animals served as objects to be studied and experimented upon in a research enterprise centering on the laboratory.

Discovering and Naming the Agent of Surra

Surra was discernable to *sarwans* as a loss of condition in camels. It was signaled by a change in appetite and dullness in the eye, and it was also detectable in the odor of an animal's urine. [5] Eventually, the animal literally wasted away. The cause of the disease was unknown, but its epidemiological pattern had long been fairly well understood in

Northwest India. It would arise in the rainy season, killing most of the camels who acquired it, and it was known to be correlated in some way with the biting flies that came with the rains. The disease developed gradually, with each stage carrying a distinct name, the last of which was surra. [6] The response of camel owners and their *sarwans* was to move the animals, if possible, away from water or marshy places, preferably into the dryer lands of the north or the arid places between the Punjab's five rivers. *Sarwans* had available to them a variety of methods and local drugs for dealing with camel afflictions, but none of these seem to have worked with surra or the conditions that preceded it. The afflicted camel would either die or, if it recovered, would be of diminished capacity, and might well go through relapses. That situation began to change with the arrival of the microscope, hypodermic needle, and eventually, the Civil Veterinary Department.

In 1880, at the end of the Second Afghan War, during which some 60, 000 military pack camels died of starvation, exhaustion, and disease, the British knew virtually nothing about surra or a variety of other diseases that affected camels and equines. Sometime in September 1880, near Dera Ismail Khan on the Indus River, veterinary surgeon Griffith Evans placed under a microscope a blood smear taken from a horse apparently afflicted with surra. In his sample Evans identified a small flagellated organism that ought not to have been there. [7] He then moved blood by hypodermic injection from the infected horse to an uninfected one and then tracked and recorded the blooming of the microbe in the blood of the second horse. After injecting a dog with the infected blood and getting the same results, Evans declared the entity he had seen to be the cause of the surra disease. Not long after, he found the same microbe in the blood of sick mules and camels (pp. 63–64).

In the report he produced on his methodology and discovery, Evans explained how he had carried out a number of autopsies on diseased horses, determining that there was no damage to internal organs. He concluded that the disease was exclusively one of the blood, and described how single and sometimes multiple microbes could be seen simultaneously attacking and ripping apart red blood cells (p. 70). The violent activity of the parasite seems to have awed Evans, and he was particularly taken by the way in which these microorganisms died off and then regenerated over time. As for the mode of transmission of the parasite into the blood of the horse, Evans thought that it was probably through food or drinking water, but he also mentioned the indigenous notion that the disease was transmitted by a large, biting fly, called by the natives *bhura dhang* ("great needle-like sting"), which swarmed around horses (p. 63). Evans recommended that more experiments be conducted to establish the mode of transmission, also arguing that until more was known about the parasite, it might be wise to retain the indigenous name for the disease — that is, surra (p. 59).

Evan's findings were printed locally in a Military Department circular in November 1880. The following year it was reprinted in installments between July 1881 and March 1882 in the *Veterinary Journal and Annals of Comparative Pathology*, the major such publication in Great Britain. In India, however, Evans received no support for his research. Neither the Surgeon-General of India, Dr. David Cunningham, nor Evans's immediate superior, Dr. Timothy Lewis, believed that the organism in question was pathogenic. Lewis based this conclusion on his study of rat blood, arguing that flagellated entities like those that Evans had found were common in "healthy" rats. Lewis added that the dog Evans had injected with surra blood was most probably suffering from "distemper" . [8] Both Cunningham and Lewis seemed to have subscribed to the theory that no microbes in blood were pathogenic, but merely the result of a body's production of an unknown chemical that generated harmless entities (p. 66). In short, Evans was accused by them of having mistaken "effect" for "cause" . [9] For George Fleming, the editor of the *Veterinary Journal* in London, however, the dismissal of Evans's research was outrageous; he saw it as just another example of the low esteem the medical profession had for veterinarians. [10]

There matters stood until, in 1885, a young veterinary surgeon named John Henry Steel identified a microbe in the blood of sick army mules in Burma as the cause of "relapsing fever" in sick mules. Steel duplicated the process of transmission from sick to healthy animals and concluded that he was dealing with the same pathogen as had been described by Evans. He made no special claims of originality, seeing himself instead as confirming Evans' work. [11] However, he also provided the first drawings of the parasite, the shape of which led him to compare it to "Spirillum Fever" .[12]

Over the next several years, research on surra was limited. A few mentions of it were published in a new journal created by John Steel, his father, Charles Steel, and Frederick Smith, all veterinary surgeons trained in London and assigned to Indian Army units. Unfortunately, the *Quarterly Journal of Veterinary Science and Army Animal Management* suspended publication in 1891 after John Henry Steel's death. [13] A few years later Alfred Lingard, the holder of the newly created position of Imperial Bacteriologist, who had been studying surra in horses, began to publish the results of his research. In this work, Lingard accepted that the cause of "horse surra" , now designated *Trypanosoma evansi*, after Evans, was pathogenic and he insisted that the path of transmission of the parasite was through the digestive system of animals. In a summary of his findings that appeared in 1894, he wrote that

> there can be little doubt that many of the outbreaks are due to the ingestion of rats' or bandicoots' excrement with corn. In the hot season, the haematozoon or its resting form in the excrement would naturally be destroyed by the continued drought; while in the rains, the moisture would be found sufficient to sustain the life of the infusoria, and enable the resting form to germinate...[14]

In other words, Lingard seemed to think of the surra parasite much as if it were a fungal spore. This led him to speculate that one possible mode of moving such spore-like parasites among populations of animals was in the draft created by fast moving trains. [15] Hence for Lingard, the expansion of the Indian railroad system had had an obvious impact on equine mortality.

Not long after Lingard published his conclusions concerning the mode of transmission of surra, David Bruce, of the Royal Medical Corps in South Africa, was working on a debilitating disease in cattle referred to by the indigenous herders of "Zululand" as *nagana*. Bruce identified what he saw under his microscope as akin to the entities described by Evans in India and Steel in Burma over a decade earlier. More importantly, he argued that the disease was moved from wild animals to healthy cattle by the Tsetse fly. His findings were published in two reports that appeared in 1895–1896, [16] and they were summarized in both the *Veterinary Journal* and in *Nature* in 1896. [17] Bruce's research was the first clear indication that the trypanosome parasite required a living agent to carry it from one mammalian host to another. It also confirmed the observations of African herders and Punjabi *sarwans* that the disease was connected to the presence of biting flies.

But the significance of Bruce's *nagana* research went beyond the set of connections he made between trypanosomes, flies, and pathology. The comparative references found in his reports, the rapidity with which his findings got into print, and the venues in which they appeared point to a new set of relationships within the domain of empire. The communication and transportation network that linked London to Zululand, India, and Burma also allowed for horizontal connections, ones that in this case put Bruce in touch with Lingard in India and made London more a clearinghouse of colonial research than a director of developments. [18] This particular trend would continue into the twentieth century. Researchers of pathogenic microbes began to think in terms of global latitudinal zones in which certain entities were specific to the "tropics" , malaria being a prime example. In the British case, the latitudinal zones covered colonies in southern and northeast Africa, South Asia, Southeast Asia and the Pacific. This geographic range meant that research being done in German and French sub-Saharan colonies, the

French colonies in North Africa and Southeast Asia, and the American colony of the Philippines were effectively made relevant to the work of British veterinary surgeons in India. [19]

India Army Reform and the Refiguring of Pack Animal Life

The trans-regional connections fashioned by steam ships, railroads, and the telegraph facilitated new information processing systems in the British empire that also relied upon low-cost print technologies and distribution networks. These networks not only made it possible for Bruce and Lingard to compare research results, but brought the events of empire — even including their research — closer to home in the European metropolitan centers. One aspect of this proximity was extensive English-language newspaper coverage of *fin de siècle* military campaigns waged in Africa and Asia by British army units. [20] There were reports, for example, of frontier uprisings in India against the British presence in the Suleiman and Hindu Kush mountains, which led to a series of punitive expeditions in 1897–1898. These mountain incursions exhausted the animal-centered Indian Army transport corps. As in the past, the army was forced to impress Punjabi camels and mules into service. But unlike in the past, serious questions were raised in Parliament about the impressment of pack animals. In answer to the queries of Lord Northbrook in the House of Lords and Mr. Edward Pickersgill in the House of Commons concerning the legal basis of forced impressment, the Secretary of State for India, Lord Hamilton, could only respond that it was "based on immemorial usage and custom" , a reply that seemed to satisfy few. [21]

Under pressure, the Government of India moved to create a much expanded transport corps that would include permanent and reserve units controlling a reliable supply of working animals. The new structure included an administrative arm made up of the Army Remount Department (ARD) and the Supply and Transport Department. On the grounds that the veterinarians of the Civil Veterinary Department had failed to produce adequate horse stock for the Indian Army, control over horse and mule-breeding operations in India was handed over to the ARD in 1903, [22] the same year the Punjab Military Animals Transport Act was implemented. This act essentially legalized impressment by establishing the colonial government's right to hire or purchase pack animals in emergencies. It also called for a census and registration of all transport animals in the province, and created Transport Registration Officers with the authority to enforce the law. [23] Meanwhile, the Supply & Transport (S&T) Department oversaw a vastly expanded pack animal establishment made up of "active duty" and "reserve" camel and mule transport units. The new organization included a provision that linked the ownership or leasing of farming plots in some of the newly created Punjab canal colonies, to the keeping of horses, camels, and mules. [24] Punjabi farmers were now required to raise animals useful for military transport. They were then organized into reserve cavalry and transport corps.

By 1905, there were approximately 2900 active duty camels in the Supply and Transport Corps. Referred to in administrative reports as government camels, they were distributed in various depots on the invasion routes leading to Kandahar and Kabul in Afghanistan. The reserve camel corps was much larger than the active duty population. It was made up of eight so-called "Silladar" regiments, [25] four Grantee camel corps (these were the units linked to the land grants), and two additional corps based at Quetta, one of which was designated the 8th and the other the Ghilzai Camel Corps. [26] Each of these units was supposed to be comprised of 1068 camels. Complementing the camel units were 21 mule corps, each of which was made up of 768 mules. In addition, there was a reserve complement of 18 corps each comprised of 192 mules. In order to function, this massive establishment of a little over 38,000 pack animals required trained transport officers, and a veterinary corps to care for both the active and reserve animals. Among other things, this meant a new kind of authority for veterinary surgeons, one embodied in the notion of "animal management" . Put simply, the new structure gave veterinarians final authority over the everyday life of pack animals as well as the right to decide which animals were fit or unfit for service. [27]

Throughout the early years of the new order, the camel units had difficulty maintaining full strength, in part because of high levels of "wastage" , an Army term for animals lost either through injury, old age, or ill-health. In the case of camels, the primary meaning of "wastage" was the number of animals that were either debilitated by or dead from surra. The S&T Department annual reports from 1901 to 1904 estimated camel wastage rates running between 30% and 40%. [28] By 1907, the department was complaining of shortages of available camels and an increase in prices for replacements, factors that led to the demobilization of one of the Silladar unit in 1908. [29] Two years later, the department reported an additional problem that the Silladar units faced — because of the expansion of irrigation systems, grazing lands available for camels were shrinking and no one appeared to have a solution to the problem. [30] Meanwhile, others were noting that where canals had been built, camel breeding was in decline. [31]

While there was little that the army could do to rein in the canal-building Public Works projects, the surra issue was more manageable within the purview of the veterinary establishment in India. The matter was placed in the hands of the Civil Veterinary Department, which then addressed it in a novel way. As camel losses mounted in 1905, the CVD took the extraordinary step of creating the position of the "Veterinary Officer Investigating Camel Diseases" . The appointment signaled that the Government of India and the Army were still very much committed to camel labor-power as part of the logistics of the Indian security regime. R. M. Montgomery, a new arrival at the Imperial Bacteriology Laboratory at Muktesar[32] in the Himalayan foothills, was given the position. For reasons not made clear in the annual report of the CVD the following year, Montgomery resigned the position. Soon after, Arnold Leese arrived from London, and proceeded to Kathgodam, near Muktesar, where he studied "Hindustani" and read up on surra. [33]

Investigating Camel Diseases

With a free hand given him by Col. Henry Pease, the Inspector-General of the CVD, Leese spent the next six years investigating camel diseases and the way of life in which they were embedded. His work as veterinary scientist and military animal manager stands as an important transition in the colonial ecology of human-animal relations. Leese conducted his research almost wholly in the field, spending most of his time near to camel grazing county in a base of operations made up mostly of tents, at Sohawa in the Jhelum district. Although the afflictions he studied included a number of ailments, such as tuberculosis, Jhoolak (skin lesions), mange, anthrax, foot-and-mouth, filariasis, and bilharziosis(intestinal parasites), Leese primarily focused his attention on surra in camels. In annual tours built around what he identified as the surra season, Leese visited all of the army's camel units, took blood samples, identified the breeding grounds of biting flies such as Tabanus and Stomoxys, and investigated the relation between weather patterns, grazing grounds, and animal health. It was grueling work, with temperatures approaching and exceeding 100 degrees daily under a blinding sun. He later recalled that he might examine the blood of up to a hundred camels in one session, "squeezing a drop out of a very slight nick in the ear of the animal on to a slide" and then setting up his microscope on the ground for viewing. [34]

Leese's first major epidemiological observation occurred at the very beginning of his appointment, while he was investigating surra among the "tonga" , that is, the fast post pony transport between the hill towns of Kathgodam and Nainithal. While making a survey of the types of flies to be found along the road connecting the two towns, he ascertained that the conventional belief held by his colleagues in the CVD that the disease spread only between October and December was wrong. The Tabanus flourished during the monsoon season from late spring into summer and had disappeared by fall. Thus, flies could only transmit trypanosomes between animals during the rainy season. It followed that the disease must incubate for a period, only showing signs of its presence in the fall and into early winter. [35] That meant that if he was going to sort out the patterns of the disease, he would have to be in the field during the time the disease was spread, the months of June to October.

In the next surra season, Leese nailed down to his satisfaction that biting flies were the vector of the disease. He selected Mohand as the site for his experiments. Surra had ravaged tonga ponies here as well, but the pony traffic had disappeared with the building of a rail line. Leese thought that by bringing both healthy and infected animals into a surra area, with no others around, he would have an ideal situation for testing. Healthy ponies were purchased locally, and camels infected with surra brought to Mohand by Leese and his assistants, Ata Mahommed and Kahan Singh. The three of them built an enclosure and covered part of it with mosquito netting. When the rains came, they put two healthy ponies under the netting (a control group kept away from flies), and four others, one of whom had been infected with surra, in open stalls. Leese eliminated grass and water as possible transmission agents by giving all the ponies the same water and local grass as portions of their feed. The same brushes, curry combs, and harnesses were used on all of the animals in order to eliminate such paraphernalia as possible transmission agents. Although some flies were present in April and May, there were no indications of the spread of the disease. When the rains arrived near the end of June, however, the situation changed dramatically.

Masses of large-size Tabanus “horse flies” , along with Haematopota and Stomoxys, appeared at the start of July. Trypanosomes showed up in the blood of the three unprotected ponies between the 4th and 17th of August, or from six to seven weeks after the rains had arrived. Flies were caught and dissected, but no trypanosomes were found in their guts. [36] Leese concluded that transmission was purely mechanical, the biting fly conveying the pathogen on the surface of its mouth-parts or proboscis and no more. [37] His findings were summarized, with extensive quotation by Pease, in the latter’s annual report on the activities of the CVD. [38] The results also appeared, with handsome illustrations of the flies, in *Indian Civil Department Memoirs*. [39] However, when no recognition of the experiment was forthcoming in the *Veterinary Journal* in London, Leese sent a letter with the results of his experiment and conclusions to the London-based *Sleeping Sickness Bulletin*, where it was printed. [40] The *Bulletin* thereupon linked Leese’s research in India to that being done by a military doctor named Friedrich Kleine in Germany’s east African colony. Kleine’s findings demonstrated that part of the life cycle of trypanosome that Bruce had identified took place inside the Tsetse fly, indicating a non-mechanical path of transmission for *nagana*. [41] The *Bulletin* noted that Kleine’s work was unknown when Leese had published his findings. Following the review of Kleine’s work, the *Bulletin* then abstracted a piece by the acting Imperial Bacteriologist in India, Maj. F. S. H Baldrey, on the “evolution” of the trypanosome associated with surra in the Tabanus fly, thus calling into question Leese’s claim that surra was transmitted mechanically. [42] In this way, the mode of transmission became a point of controversy in India between Leese and the Muktesar establishment, especially because, morphologically, the African and Indian trypanosomes appeared to be identical.

I emphasize the publication trail here not only to demonstrate the venues then available for the circulation of medical and militarily useful information, but to make two other points as well. The first has to do with comparative research on trypanosomiasis in European colonies. In this case, Kleine's studies in Africa built on those of Bruce in Zululand, while confirming that biting flies were the vector of trypanosome transmission in animals and humans. [43] Second, although Leese himself admitted there were a few questions still to be answered about transmission, he had demonstrated through live animal experiments how an epidemiological approach to the disease could resolve long-standing speculation about how trypanosomes infected horses and pack animals.

Over the next four years, Leese refined these initial conclusions by expanding his range of observations across the Punjab and into Baluchistan and the Sind. As he made annual tours of the frontier and Silladar units to collect data on the scale of infectious animal diseases, Leese was also able to establish a relation between water, be it irrigation canals or monsoon rains, and the life cycles of biting flies. In the case of Tabanus, the fly's larvae were found on the leaves of plants overhanging the canals and in the areas where they overflowed in flood season. [44] This led Leese to advocate what he thought of as a simple preventive measure. Inspectors should destroy larvae when found and then move camels, if possible, into grazing grounds free of canals and pooled rain water. As an example of how effective such a procedure could be, he pointed to the 67th Punjabi Camel Corp. During his tour of this unit, he learned that they regularly moved their camels into surra-free grazing areas during the rains; as a result they had the lowest mortality rate among camel transport units. [45] Having established to his satisfaction the relation between water and biting flies, Leese also used his field tours to identify naturally occurring surra-havens. [46]

Another purpose of these tours was to place the facts he was producing in the hands of unit commanders in an effort to enroll them and their *sarwans* into his research design. Part of his strategy to accomplish this goal was to provide microscopes to *sarwans* and encourage them to examine camel blood. Since he knew that surra was present in virtually all of the units, his idea was to show the *sarwans* the actual agent of the disease and thereby convince them to become part of his surveillance and disease management scheme. This strategy no doubt involved an effort to establish a firm statistical base for the disease. But the spread of the microscope was also designed to demonstrate to the camel corps commandants and natives the superiority of this visual technology in identifying infected animals expeditiously. [47] In 1909, Leese was happy to report that the commandants of the units and their *sarwans* had developed confidence in the microscope and had begun culling infected animals from their units. [48]

By emphasizing to camel corps commanders the relationship between biting flies, the rainy season, and likely fly breeding grounds, Leese was able to persuade many commandants that prevention was the key to managing the disease. Prevention involved a set of practices designed to reduce risk and hence, the incidence of infection. But to work, this mode of animal management, as Leese both argued and demonstrated, required the vigilant and regular inspection of all of the camel transport units by the camel specialist, the full cooperation of camel corps commandants and their *sarwans*, and the availability of ample grazing land onto which camels could be moved.

At the same time, however, Leese was keenly aware that his efforts to identify safe grazing grounds were being undercut by canal construction. He noted at one point that a good area in the Chenab Colony should be used immediately for grazing because canal expansion was planned there. If the canal system was to be further expanded in the Punjab, he repeatedly argued, then more government-owned rakhs (wilderness areas) free of surra, but presently closed to pastoralists and the camel corps, needed to be made available for the transport camel units. [49] Otherwise, it would be impossible to reduce camel and equine losses to the disease.

Fieldwork also led Leese to make a number of other recommendations. For example, he argued that the temporal tenure of camel unit commandants was too short. As he put it, just when an officer learned about camels and developed a bond with his *sarwans*, his knowledge and experience was lost through transfer. [50] He also learned from *sarwans* that keeping camels in good condition involved attention to diet. Camels required a mixture of wild plants consumed while grazing and missa bhusa or pea straw when not. [51] Such a mixture aided ruminant digestion. Leese was also concerned about another possible concentration of surra, periodic animal fairs that took place in different parts of the Punjab. He advised that any camels purchased at these fairs should only be bought outside of the surra season, accompanied by microscopic blood tests. [52] Moreover, since only one-third of Silladar camels were on active duty at any time, Leese thought that the other two thirds ought to have their grazing areas inspected for biting flies and their blood checked for the parasite. [53]

While all of Leese's ideas made sense from the point of view of animal management, disease control, and transport unit readiness, implementing them was another matter. It would have required the army to commit vast new resources to provide the personnel to carry out the surveillance and policing that Leese was calling for. But if his ideas about improving transport were probably unrealistic, the epidemiological and animal management approach to surra that he had pioneered, when combined with the cooperation of unit commanders and ruthless culling, began to show positive results by 1911. The CVD reported that mortality in the eight Silladar Camel Corps had fallen from over thirty percent to 13.3 percent on an average annual strength of 8, 230 animals. This was deemed "satisfactory". [54]

Once he had managed to reform camel transport unit practices and increased awareness of the nature of the disease, Leese then turned his attention to experimenting with possible cures for surra. This research would eventually lead to clashes between Leese and veterinarians at the Imperial Bacteriological Laboratory in Muktesar, where surra in horses had been under investigation since Lingard established the laboratory in 1895. [55] The new director, Maj. J. E. D. Holmes, who had taken over when Lingard retired in 1908, was no less keen to find a cure. Like the veterinarians at the laboratory, Leese began to experiment with drugs such as Atoxyl, an organic arsenic compound manufactured by a German firm; Soamin, a version of Atoxyl produced by Burroughs Welcome; [56] Tartar emetic (Antimony potassium tartrate), and Sodium Arsenate. [57]

It is clear from Leese's initial foray into chemotherapy that there were a number of problems related to using this mixture of poisonous compounds on camels. First and foremost was the symptomatic pattern of the disease. Recall the initial identification of the disease by Evans and Steel as "relapsing fever" . During the paroxysms of fever, trypanosomes were abundant in the peripheral blood vessels of the animal's body. When the trypanosomes disappeared from the peripheral blood, the body temperature returned to normal. Where the trypanosomes went and how they later regenerated was unknown. Without treatment, a certain number of animals would survive the initial attack and subsequent relapse incidents, eventually appearing surra free. Given this pattern, the issue was how to distinguish those who spontaneously recovered from those animals who appeared to recover with chemotherapy. A second question involved time — how long should researchers continue testing blood until an animal could be declared cured?

A third kind of issue concerned what was achieved through live-animal experiments involving chemotherapy even when they appeared successful. By 1911 both Leese and the researchers at Muktesar had demonstrated that a combination of Atoxyl, Tartar emetic, and Sodium Arsenate spread over several days in alternating and carefully measured doses seemed to eliminate the parasite from some animals infected with *T. evansi*. [58] However encouraging this might have been for the veterinarians and the army transport corps, chemotherapeutic successes generated their own problems. Leese spelled these out in his annual report. The most obvious problem was that some animals died from arsenic or antimony poisoning, while others sustained internal damage. Animals could develop lung-abscesses, heart dilation, and severe constipation in which the third and fourth stomachs of camels became impacted with dried food. [59] Leese estimated that about half of the animals that arrived at his base in Sohawa were already in such terrible shape that arsenic treatment insured their demise. But even those that survived were so debilitated that they required an extended recovery period, with attendant costs to the army for their care.

Moreover, the animal populations that Leese and the Muktesar laboratory establishment experimented with were distinctly different. In Leese's case, he treated camels with this drug cocktail, while the Imperial Bacteriologist, J. E. D. Holmes, carried out his tests on horses. No one knew, for example, the significance of the body weight of animals beyond the general assumption that larger animals could manage larger chemotherapeutic doses. Nor was it clear how the physiologies of different animal species might interact with the parasites, or why some animals, like Lewis's rats, seemed immune, while others died soon after exposure. There was also an absence of protocols in place to continue testing animals declared surra-free once they were "cured" by chemotherapy. Instead, arbitrary time periods were selected (250 to 350 days) and success declared if the animal's blood showed no return of trypanosomes. At the same time, veterinarians agreed that a cured animal could be reinfected in another surra season.

Finally, the drug regime posed one other critical problem. Diagnosis of surra through blood tests could find the pathogen present in what appeared to be perfectly healthy camels. Not surprisingly, some *sarwans* in the reserve units were incredulous about allowing their apparently healthy camels to be treated by a method that could possibly kill them, leaving the owner responsible for their replacement. At one point, W. S. Hamilton, the successor of Henry Pease as Inspector-General of CVD, suggested that the government needed to offer suitable compensation to owners in order to encourage treatment. [60] Otherwise, there was the real danger of unwittingly retaining infected animals in the inactive reserve, where microscopic testing was less common. Whether or not his suggestion was taken up is unclear in the records.

Such concerns, however, seemed to pale in the face of the pressure within the transport system to maintain mobilization-ready units in the camel corps. In its annual report of 1915, for example, the Punjab Veterinary Department noted that in spite of Leese's ability to reduce the death rate and the use of chemotherapy, surra deaths among camels had gone from twenty-five in 1912–13 to two-hundred eighteen in the current year, requiring a substantial number of replacements. The report added that "until the movement of animals can be restricted (which is a very big problem), the disease is bound to spread" . [61] The report was speaking not only of traditional patterns of pastoralist seasonal migration into and out of the Punjab "wastelands" , but also of camel owners from outside of the Punjab, who had moved animals in to meet the demands of the military transport market there. The perceived problem was that these camel owners might see the demand as a way to get rid of infected or aged animals.

At the same time however, movement of camels was, in some areas, already restricted by the Forestry Department, which controlled most of government controlled rakhs (wild preserves). [62] As Leese had noted earlier, while good grazing grounds existed outside of surra zones, camel herders were generally barred from the ones that were part of rakhs. Yet, as grazing land shrank because of canal expansion and rakh restrictions remained in place, pastoralists and the reserve camel corps units had to move far afield from their usual grazing grounds. In other words, at the same time as Leese was calling for much tighter control over camel life in order to contain surra, the policies of other government units such as the Forestry and Public Works Departments were making it difficult for the CVD and the Army Remount Department to manage the camel population. Moreover, other than the Transport Registration Officers, whose purview was limited to their own district, there were few resources at hand to keep track of the movement of camelmen seeking surra-free grazing lands, let alone conduct epidemiological surveys of animal diseases.

On the supply side of the issue, camel reservists and camel land grant holders faced a number of difficulties in maintaining their arrangement with the government. The annual reports of the Chenab Canal Colony, where the bulk of the camel land grants were located, indicated a disturbing pattern. The camel grantees and the units they were attached to were regularly short of camels, primarily because each year so many of the animals were lost to surra. Few grantees could afford to replace their camels, however, and they were forced to take government loans, or *takavi*, in order to do so. At the same time, the colony reports indicated the shrinkage of suitable grazing land for these animals. The result was that grantees, already financially strapped, had to purchase animal feed, a situation that, as one report put it, made it impossible for them to make a living. [63] These problems were repeated in the reports of subsequent years, with shortages in the four camel grantee corps reaching 308 in 1910. [64] By 1914, the Chenab report suggested that there was a general decline in the quality of existing camels as a result of stall-feeding and lack of grazing grounds. [65] Concerned even earlier that the camel grantees were being forced into untenable positions, the Punjab Government appointed a Camel Browsing Committee to investigate matters. Their report of December 1911 recommended that because of the increased shrinkage of grazing lands, the grantee camel units ought to be decommissioned. [66] The war in Europe intervened, however, before any action could be taken.

Tensions within the Civil Veterinary Department

Leese's empirical fieldwork between 1908 and 1912 not only produced new knowledge about surra and other camel diseases, but also demonstrated that a good deal of earlier speculation and laboratory experimentation about surra was either incorrect or suspect. Problems, where they existed, were more often than not in the work and publications of the imperial bacteriologists, beginning with Arthur Lingard and his notions about the oral ingestion mechanism of surra transmission in the 1890s. Leese had corrected this notion with his work on the mechanical transmission of surra through biting flies.

Matters might have rested there if it were not for the publication of Baldrey's article on the "evolution" of the trypanosome within the body of the Tabanus fly in the same issue of the *Sleeping Sickness Bulletin* in which Leese's letter on surra appeared. Apparently influenced by Kleine's work on the developmental cycle of *T. brucei* in the Tsetse fly, Baldrey claimed to have found a similar process in the Tabanus, thus casting doubt on Leese's model of mechanical transmission. [67] More importantly, perhaps, in comparison to the less well authenticated Arnold Leese, Baldrey could claim a certain authority through the substantial number of acronyms with his name (Maj., F. R. I. P. H., F. R. C. V. S., and F. R. G. S.), [68] his diverse publication record, and a professorship in Sanitary Science at the Punjab Veterinary College.

Leese may have seen the article or knew of Baldrey's work, but rather than write the journal in which it appeared, he sent the letter mentioned above to the editor of the *Sleeping Sickness Bulletin*. As we saw, Leese made clear that he had also dissected flies, found no changes in the trypanosomes in the fly's gut, and concluded that transmission was mechanical. But he went further; he reinforced his conclusion by citing the results of his research on a camel surra epidemic in the Bikanir desert in 1910. The Bikanir was an area far removed from canal irrigation and Leese had found neither Tabanus nor Stomoxys flies present. Instead, he discovered a new vector, the *Lyprerosia minuta*, which had apparently been stimulated into a massive bloom by unusually heavy rains during the monsoon season of 1908. [69] Leese thought it completely implausible that *T. evansi* could carry out part of its life cycle in three or more fly species. Moreover, did it really matter, he wondered, whether part of the trypanosome's life-cycle occurred in a fly? Wouldn't research time be better spent experimenting on treatments for the disease, instead of dissecting flies? [70]

The following year, the *Journal of Tropical Veterinary Science* published a much expanded version of Leese's letter. In it, Leese began by noting that Kleine's work on the life-cycle of the trypanosome in the African Tsetse fly seemed to have "cast grave doubt" on the direct spread of surra and that "High authorities" in India now believed that development of the parasite in an invertebrate host in India was "very probable." [71] After repeating his Bikanir findings (21–23), he proceeded to refute Baldrey's laboratory research, emphasizing that the whole process of biting flies moving surra can be "seen in the Monsoon season in any Tabanus-afflicted grazing ground" (24) — that is, it was clearly visible in the field! Or as he put it on another occasion, any observer of camels in a heavily infested Tabanus region, can see the flies themselves experimenting with "direct inoculation." [72] But he didn't stop there — he then turned to the publications of the current Imperial Bacteriologist, J. D. E. Holmes, who in 1906 had expressed doubts about the mechanical transmission of surra or of the possibility that an animal infected with surra could be imported into an area and become the source of the spread of the disease (25). Leese had demonstrated just the opposite in his very first surra experiments, apparently without any acknowledgement from Holmes.

Leese then moved the discussion into another realm. Infected animals that survived the initial onslaught of the disease, he insisted, carried surra from one season to the next, thus becoming a reservoir for the next cycle of biting flies to spread the disease (28–30). With this notion of camel as reservoir, Leese then speculated on how surra got to India. He thought that at some earlier date camels had entered the African Tsetse belt region in the normal course of trade between northern and southern African groups. Camels had been bitten and infected with nagana or bovine trypanosomiasis. Because the animal could carry the trypanosomes for up to three years or longer in its system, trypanosomes moved north and east, following trade routes and finding new vectors along the way (31–32). This deduction, in part based on studies of trypanosomiasis in Africa, and with additional support from comparative genetic studies of *T. evansi* and various Tsetse belt trypanosomes, would become the standard interpretation of the blood parasite's globalization. [73]

His response to Baldrey and to laboratory science was Leese's final publication in India on surra. In 1912, he went on an extended leave. When he returned, he was informed that his position as camel specialist was rescinded and that the government wanted him to take up the study of the diseases of the elephant, an animal that had more or less been eliminated from army transport service. Leese says little about the circumstances of this decision in his memoir, and there seems to be nothing in the archival records indicating that he was being rifted by "higher authorities" , but his open challenge to the work of his superiors could not have stood well with many. In his memoir, he only mentions that he thought it too late in life to begin the study of another highly unusual animal. Given his admitted disdain for military discipline and the "social conventions" of "station life" in India, as well as his appreciation of the willingness of his first superior, Henry Pease, to give him a free hand in his earlier investigations, [74] it also seems possible that he felt a modicum of resentment toward the army officer veterinary surgeons who dominated the Civil Veterinary Corps and the Muktesar Laboratory. He may also have bridled at the fact that, other than Pease, few of the higher authorities in the CVD seemed to have acknowledged his contributions.

But there was also something else involved in Leese's outbursts in the biting-flies papers. It highlights tensions between a field-work oriented epidemiological approaches to disease and one focused on laboratory science. Recall Leese's comment on the life-cycle of the trypanosome. From his point of view, research time spent on trying to determine if part of its development occurred in the fly helped little in solving the bigger problem of keeping the camel transport units near their full compliment. Knowledge of the dynamics of the surra season, of surra and non-surra territorial zones, and of the costs and benefits of drug cures for the disease made sense in relation to the primary needs of the military. And those needs were for professional animal management in the service of military transport and cavalry units, and their readiness for mobilization.

For CVD laboratory researchers, on the other hand, even though most of them were military officers, the desire for comprehensive knowledge of the workings of an organism, like the trypanosome, was taken for granted. Their will to knowledge demanded that the morphology, life-cycle, and mechanisms of transmission of pathogens be intimately understood, particularly at the biological level. For those working at Muktesar, the human-animal-plant-climate complex in which camels were embedded was immaterial to the questions on which their microbiological endeavors were centered. They tended to ask less ecologically-oriented questions. Instead the ones they asked looked something like this: Were other animals susceptible to surra in the same way horses and camels were? Was there one kind of surra, or as some research suggested, a dog, horse, camel, and bovine variety? How could animal models best be used to answer such questions? Arsenic compounds showed some promise as a treatment, but were also debilitating to the host. If arsenic compounds were to be employed, in what proportion or combination with other drugs would they be most effective? What was the exact relationship of contiguity between vectors of transmission and the parasite's animal host, and how would such knowledge help in the battle against the disease?

These were the sorts of questions that animated Muktesar laboratory scientists and structured the tropical veterinary research regime in the early twentieth century. In it, pack animals took on a new form of service as laboratory animals. Along with horses, mules and a menagerie of live experimental animals (e.g., guinea pigs, goats, dogs), camels were impressed once again as forced laborers, even outside of military campaigns and frontier warfare. As symptom complexes and living laboratories, they were the objects of the scrutinizing eye of the veterinary surgeon as scientist. They became laboratory natural resources as providers of blood. Their blood, in turn, entered a structured epistemological complex, an assemblage, made up of microscopes, glass slides, photographic equipment, veterinary scientists and native lab technicians, statistical forms of representation, and entries in professional veterinary journals in and outside of India. Mules and camels were re-embodied as blood smears, temperature graphs, and photographic images of diseased and experimental animals. They became receptors for carefully timed and calculated doses of poisonous synthetic drug mixtures, each of which was numbered, like the animal, for evaluation, comparison, and eventual archiving. Animals that had been "macroscopically complex" , or masses of uncontrollable variables, were translated, through the laboratory regime, into "docile specimens" . [75]

At the same time, the laboratory complex was perfectly capable of appropriating and correcting Leese's field research and critical insights. Such was the case with Leese's successor, H. E. Cross. An assistant bacteriologist at Muktesar beginning in 1906, Cross worked with Holmes on early experiments with Atoxyl, Tartar Emetic and other drugs in treating surra in horses. [76] In his first annual report after becoming camel specialist, Cross set about amending Leese. He pointed out that the latter's cure for surra in camels was no more than a modification of the drug regime that he and Holmes had pioneered, and he argued that Leese's results were based on too small a sample size ("only" nine animals). [77] The next year, Cross called into question Leese's insistence that surra was transmitted mechanically. He referenced Tsetse fly research and then argued that if transmission were mechanical, far more animals would be infected than actually were. This observation led him to question whether the Tabanus fly species was as big a culprit in spreading surra as Leese had suggested. [78] Cross's corrections were, in a sense, also a research agenda. Over the next decade, he became fixated on finding the right dosage and timing for exclusively using Tartar

Emetic as a treatment for surra, while formulating a seemingly endless series of experiments to establish the feeding habits of multiple kinds of biting flies. [79]

These examples of differences between a field-oriented veterinarian like Leese and laboratory-oriented one like Cross might be read as a personality clash or an example of the rivalries that were common to the scientific production of truth, in the metropole as in the colonial field station. [80] They might also be understood as exemplifying the difference highlighted by Michael Worboys as the "essential tension" between two forms of scientific knowledge. The kind of science that Leese was producing was gained through research that was "problem" - oriented or "subject-based" , creating knowledge that was "particular" or "empirical" . Muktesar scientists, in contrast, oriented to the production of "universal" or theoretical knowledge, produced it through "distinguishable cognitive structures and technical resources" . [81] Cross's critique of Leese's seat-of-the-pants methodology and conclusions drawn from small sample sizes seems to exemplify a contrast between a universalistic vs. particularistic research concerns. To Leese's credit, however, his approach, particularly that involving the identification of surra-free territories, drew on and left a space for indigenous knowledge. Had he not followed a problem-based approach that had him listening to the *sarwans*, he might not have understood existing local patterns of seasonal animal movement. The laboratory scientists, by contrast, had little use for pre-microbiological understandings of animal diseases.

On the other hand, Leese and Cross also shared certain assumptions. Both put a great deal of emphasis on the techno-scientific apparatus surrounding the microscope and hypodermic needle; both seemed to understand the importance of publication to any claims they might make about scientific research and practice; and both were clearly committed to laboratory-made synthetic drugs and vaccines. These commitments pointed to changes evident in veterinary medicine in the early twentieth-century.

In India, British veterinarians had, for much of the nineteenth-century, tended to rely on a small number of basic drugs they brought from England, while also drawing on large numbers of local herbals that were part of the Indian pharmacopeia. [82] And while they might disdain Indian formulary for its ignorance of the chemical properties of medicinal herbals, veterinarians found that they had no choice but to rely on things available in the bazaar, both because of their low cost and the absence of alternatives.

By the 1890s, however, veterinarians operating in the colonial tropics increasingly had access to new kinds of drugs. Chemists, "no longer… at the mercy of Nature," [83] developed methods to synthesize organic and inorganic compounds like Atoxyl and Tartar Emetic in their chemistry laboratories. And, some medical doctors and veterinarians found that they had a captive population for working out the efficacy of these chemical compounds to treat a variety of afflictions. Parasitic infections, for which there were no clear cures or protocols of treatment, were ideal targets against which to direct man-made substances such as Atoxyl. [84]

Not long after Cross published his final study of the use of Tartar Emetic to cure surra in camels, Capt. R. H. Knowles, of the Royal

Army Veterinary Corps and a Veterinary Research Officer for the Sudan Government, called into question Cross's approach. In addition to being "somewhat prolonged" , Knowles argued, Cross's method was problematic because trypanosomes, more often than not, reappeared "after a variable period" . Knowles then provided data on his tests with a new pharmaceutical drug, "Bayer 205" . This was a urea-based compound invented in the laboratories of the German drug manufacturer Bayer in 1916. Later known as Naganol, the compound not only cured surra in Knowles's tests, but didn't seem to have the negative physiological effects on camels common to treatments with Atoxyl and Tartar Emetic.

The arrival of Naganol in Africa, and soon after in India, also pointed to something novel. That British veterinarians were field-testing the product of a German pharmaceutical company signaled the further development of a transnational economy of manufactured drug compounds. [85] And the linkage between field, manufacturer and army veterinary practice was institutionalized almost immediately. While the section on surra to be found in the third edition of the *Handbook on Contagious and Infectious Diseases of Animals* (1929) gave pride of place to Cross's Tartar Emetic treatment, it also advised the use of 4 grams of Naganol and 3 grams of Tartar Emetic, given in a single intravenous injection. The simplicity of the method, the text tells us, is that it will "obviate the evacuation of Surra cases for treatment to [Lines of Communication] and Base Camel Hospitals during campaigns, as camels in good condition can be cured and returned to duty within a few days" ... rather than months. [86] Put simply, Naganol promised to finally solve the logistical problem surra had created during the frontier campaigns that relied on the labor of camels and equines.

Aftermath

By the time the single dose treatment of Naganol arrived on the scene, however, substantial change had begun in the army transport service. In 1921, due to the shortage of grazing land, peasant indebtedness, and surra, the camel land grant leases and the grantee system were allowed to lapse in the Chenab colony. Peasants were permitted to purchase the land under colony tenure arrangements. [87] Soon after the dissolution of the camel grantee units, the transport corps began a gradual shift from animal to mechanical transport, and the mule and camel units were downsized or eliminated. Recall that in 1905, the total establishment of active and reserve camel and mule units was some 38, 000 animals (18, 500 camels, 19, 500 mules). In a series of cuts and reconfigurations, the numbers had been reduced to 12, 576 mules and 4, 704 camels by 1930. [88] And even this number might have been lower if it were not for the India Army's continued engagements in mountains accessible only to camel and mule transport.

Ironically, much of the research conducted on surra by Steel, Leese, Cross, Holmes and others was relegated to the dustbin of history. The fourth edition of the *Handbook on Contagious and Infectious Diseases of Animals* (1936) completely eliminated Cross's Tartar Emetic treatment, noting that like Atoxyl, the former compound only killed trypanosomes in the peripheral circulation of animals, while survivors took "refuge" in the "cerebro-spinal canal" . It then indicated that "4 grammes" of Nagonal had proven entirely successful in curing surra in the Sudan and in the Army of India. [89] As the Principal of the Punjab Veterinary College in Lahore put it in 1934, the records of the treatment of surra with Atoxyl and Tartar Emetic "have come to be regarded as so many 'dead' chapters in the history of chemotherapeutic endeavors..." On the other hand, as the Principal also noted, surra was alive and well, occasionally assuming "terrifying proportions" in the Punjab. [90] A few years after the Principal's statement was published, the parasite surprised everyone by overcoming the historically resistant bovine population of India and killing huge numbers of cattle. [91] Whether this development was related to the experimental moving of blood between animal species is unknown, but at least one recent group of researchers have noted the "incredible plasticity of this amazing parasite" , [92] which may explain why surra remains an active agent in the Punjab. The difference now is that it is a problem for the post-colonial development state to manage. [93]

Today, vast stretches of Punjab's "wasteland" , formerly the browsing grounds of the region's camel population, remains agricultural land, while the lives of the region's pastoralists and camel breeders are more a curiosity than a significant part of the political economy of the region. Now development officials make arguments for the economic value of the camel's milk, meat, and other products, rather than its value as a transport animal. [94] Meanwhile, along the dense network of Punjab irrigation canals, those material monuments to colonial engineering expertise and imperial hubris, numerous species of biting flies continue to find ideal breeding grounds. These include not only the transmitters of surra, but also the anopheles mosquito, the vector of malaria. [95]

Notes

[1] Timothy Mitchell, *Rule of Experts* (Berkeley: University of California Press, 2002).

[2] David Gilmartin provides a detailed analysis of these changes in the Punjab beginning in the last two decades of the nineteenth century and their impact on the local population; see *Blood and Water* (Berkeley: University of California Press, 2015).

[3] See David Gilmartin, "Scientific Empire and Imperial Science: Colonialism and Irrigation Technology in the Indus Basin" , *Journal of Asian Studies* 53. 4: 1127–1149 and Imran Ali, *The Punjab Under Imperialism, 1885–1947* (Princeton: Princeton University Press, 1988).

[4] See James Hevia, *The Imperial Security State* (Cambridge: Cambridge University Press, 2012).

[5] On the capacity of native camel men to identify the disease, see Henry T. Pease, "Tibarsa Surra: Trypanosomiasis in the Camel" , *Journal of Tropical Veterinary Science* 1. 1 (1906): 75. Pease added that he was unable to do the same without the aid of a microscope.

[6] In his report on surra (see below), Griffith Evans, who discovered the pathogen in question, provided native testimony on the stages of the disease. See his "Report on 'Surra' Disease," which originally appeared in a *Punjab Government, Military Department* circular dated 13 November 1880 (No. 493–446), 65.

[7] I draw on Evans's "Report on 'Surra' Disease" , and cite the page numbers from it in text.

[8] Griffith Evans, "Autobiographical Memoir" , *Annals of Tropical Medicine and Parasitology* 12. 1 (July 1918): 14. The conclusions of Cunningham and Lewis concerning Evans' research may help account for the absence of any mention of surra under the camel entry by George Watt in his *The Dictionary of Economic Products of India* (Calcutta: Superintendent of Government Printing, 1885–1896), 10 vols. Also see Watt's pamphlet *The Camel*, which was published in 1887.

[9] This was the conclusion drawn by Sir Frederick Smith from his evaluation of the Cunningham and Lewis assessment of Evans's findings. Smith, a veterinarian and historian of the profession in Great Britain, thought them both wrong and unnecessarily hostile to Evans. See his *A History of the Royal Army Veterinary Corps, 1796–1919* (London: Bailliere, Tindall and Cox, 1927), 174.

[10] George Fleming, "An Unwarrantable Intrusion" , *Veterinary Journal* 13 (July

1881): 133.

[11] India Office Records, London (hereafter, IOR) V/27/541/15: "Report of Veterinary Surgeon J. H. Steel, A. V. D., on his Investigation Into an Obscure and Fatal Disease among Transport Mules in British Burma, which he found to be a Fever of the Relapsing Type, and probably identical with the disorder first described by Dr. Griffiths Evans under the name "Surra" , in a Report (herewith reprinted) published by the Punjab Government, Military Department, No, 439–4467, of 3rd December 1880 – *vide* the *Veterinary Journal* (London), 1881–1882.

[12] John Henry Steel, "On Relapsing Fever of Equines" , *Veterinary Journal* 22 (1886): 166–174. Spirillum fever, which turned out to be a miss-classification, is caused by a tick-transmitted bacteria. According to Cecil Hoare, Steel named the pathogen *Spirochaete evansi*; see *The Trypanosomes of Mammals* (Oxford and Edinburgh: Blackwell Scientific Publications, 1972), 555.

[13] J. H. Steel introduced a column in the journal entitled "Cameline Pathology" , which ran over several issues. In 1890, he collected these pieces together in *A Manual of the Diseases of the Camel and His Management and Uses* (Madras: Lawrence Asylum Press, 1890). His summary of surra (pp. 46–47) indicates that little research on it had been done following the identification of the trypanosome as the cause of the disease.

[14] Alfred Lingard, *Summary of Further Report on Surra* (Bombay: Government Central Press, 1894), 8. Infusoria was a term used to encompass all microscopic life in fresh water.

[15] Alfred Lingard, *Report on Horse Surra* (Bombay: Government Central Press, 1893), 10–11. Also see A. Lingard, "Horse Surra" , *Civil Veterinary Department Ledger Series No. IV* (1894): 14, in IOR/V/25/541/2.

[16] David Bruce, *Preliminary Report on the Tsetse Fly Disease or Nagana, in Zululand* (Durban: Bennett and Davis, 1895) and *Further Report on the Tsetse Fly Disease or Nagana, in Zululand* (London: Harrison and Sons, 1896).

[17] "The Tsetse Fly Disease" , *Nature* 53: 1381 (Apr. 1896), 566–567. The article indicated that the nagana microbe was "allied to, if not actually identical with, *Trypanosoma evansi*, the haematozoon of "Horse Surra" (567).

[18] On the Bruce-Lingard correspondence, see Bruce, *Further Report*, 17.

[19] The tropical colonial network and the scale of research on trypanosomes can be discerned, for example, in C. A. Thimm, comp., *Bibliography of Trypanosomiasis* (London: Sleeping Sickness Bureau, 1909). Also see the sources cited in W. E. Musgrave and Moses Clegg, *Trypanosoma and Trypanosomiasis with Special Reference to Surra in the Philippine Islands* (Manila: Bureau of Public Printing, 1903) and A. Laveran and F. Mesnil,

Trypanosomes and Trypanosomiasis. Trans. David Nabarro (London: Baillière, Tindall and Cox, 1907).

[20] For a discussion of media coverage of Britain's "small wars" , see James Hevia, *The Imperial Security State* (Cambridge: Cambridge University Press, 2012), 232–248.

[21] See India Office Records (hereafter, IOR) L/MIL/7/6683. The file includes the response of Lord Hamilton, Secretary of State for India, to the question concerning the legality of impressment.

[22] IOR/L/MIL/17 T. N. 544: "Report of the Horse and Mule Breeding Commission assembled under the Orders of the Government of India, 1900–1901," 35, 61–62. On the other hand, the British government finally moved in 1903 to alter the status of army veterinary surgeons by sanctioning the creation of the Royal Army Veterinary Corps; see F. Smith, *History*, 209–210.

[23] See Government of India, Legislative Department, *The Punjab and North-west Code,* 3rd ed. (Calcutta: Superintendent of Government Printing, 1903), 518–527.

[24] The outlines of the new structure can be found in the report from the Viceroy, Lord Curzon, and his council, to the Secretary of State for India dated 13 September 1900. The reform was based on the report of the Transport Committee of 1898. See respectively IOR/L/MIL/7/6702 and IOR/L/MIL/7/6688. On camel service grants, see *Gazetteer of the Chenab Colony, 1904* (Lahore: "Civil and Military" Gazette Press, 1905), liii-lviii.

[25] The irregular cavalrymen provided their own horses and armaments; see *Hobson-Jobson*, 386.

[26] See *Quarterly Indian Army List, July 1908* (Calcutta: Office of the Superintendent for Government Printing, 1908), 405a-f. In a report of 1909, Arnold Leese, the CVD camel specialist (see below) indicated that he inspected additional camel units attached to the Corps of Guides, the 67th Punjabis, the 58th Rifles, the 14th Sikhs & 84th Punjabis, and Supply & Transport units at Bannu (the Kabul line) and Dera Ismail Khan (the Kandahar line); see *Report of the Veterinary Officer investigating Camel Diseases for the Year Ending 31st March 1909* (Simla: Government Central Branch Press, 1909), 3.

[27] See *Field Regulations, India Provisional Edition]* (Calcutta: Superintendent of Government Printing, India, 1906), 16–17, where the authority of veterinary surgeons is spelled out.

[28] The reports are bundled together in IOR/L/MIL/6700. See *Administration Report of Supply & Transport Corps for the official year 1899–1900* (hereafter ADS&T; Calcutta: Superintendent of Government Printing, India, 1901), 20;

ADS&T (Simla: Government Central Printing Office, 1902), 16; and ADS&T (1903), 17. For unclear reasons, subsequent reports drop percentages and merely claim that wastage figures are either higher or lower than the previous year.

[29] ADS&T (1909), 4. Another rational given for the demobilization decision was that it saved money. The report also claimed that registration efforts were so productive that sufficient camels would be forthcoming for hire upon mobilization. Reports of shortages of replacement animals for the Silladar and Grantee units contradicts this assumption, however.

[30] ADS&T (1907), 4; ADS&T (1909), 4, 6; and ARS&T (1910), 10. See IOR/L/MIL/6712.

[31] *Annual Report of the Veterinary Officer*, 1910, 5.

[32] Originally the laboratory was in Puna, but Lingard lobbied the government successfully to have it moved away from the heat, dust and filth of the lowlands into the mountains. On Lingard's role in the establishment of the Muktesar laboratory, see Pratik Chakrabarti, *Bacteriology in Colonial India: Laboratory Science and the Tropics* (Rochester: University of Rochester Press, 2012), 62–65.

[33] Arnold Leese, *Out of Step: Events in Two Lives of an Anti-Jewish Camel Doctor* (Guilford, England, 1951), 16.

[34] Leese, *Out of Step*, 17.

[35] Leese, *Out of Step*, 16.

[36] See *Sleeping Sickness Bulletin* 3. 26 (1911): 369.

[37] "Camel Surra", 9–11, in *Report of the Veterinary Officer Investigating Camel Diseases for the Year Ending 31st March* 1909. (Simla: Government Central Branch Press, 1909). Leese also published a version of this report in "Experiments Regarding the Natural Transmission of Surra Carried Out in Mohand in 1908", *Journal of Tropical Veterinary Medicine* 4.2 (1909): 107–132. He also briefly outlined the experiment in *Out of Step* (19).

[38] See *Annual Administrative Report of the Civil Veterinary Department, 1908–1909* (Simla: Government Central Branch Press, 1909), 14–16.

[39] See *Indian Civil Department Memoirs* 2 (1909): 1–26.

[40] *Sleeping Sickness Bulletin* 3. 26 (1911): 367–369.

[41] Kleine's work appeared in "Sleeping Sickness Investigations in German East Africa", *Sleeping Sickness Bulletin* 3. 26 (1911): 165–175.

[42] See *Sleeping Sickness Bulletin* 3. 30 (1911): 369–373. For Baldrey's article, see "The Evolution of Trypanosoma evansi through the Fly", *Journal of Tropical Veterinary Science* (hereafter cited as JTVS) 6. 3 (1911): 271–282.

[43] Whether the African trypanosome, *T. brucei*, and *T. evansi* were the same entity was another matter. Later research would indicate that in fact the latter was a descendant of the former, but was distinctly different because it could only be transmitted mechanically; that is, it would not thrive in the gut of a fly. See the discussion in Cecil Hoare on the history of the discovery of this difference in *Trypanosomes of Mammals*, 382–383.

[44] *Report of the Veterinary Officer*, 1909: 6 and 1910: 5, 7.

[45] *Report of the Veterinary Officer*, 1909: 8.

[46] *Report of the Veterinary Officer*, 1910: 6.

[47] *Report of the Veterinary Officer*, 1909: 1, 5–6.

[48] *Report of the Veterinary Officer*, 1909: 6.

[49] On grazing issues, see *Report of the Veterinary Officer*, 1909: 4 and 1910: 2, 5.

[50] *Report of the Veterinary Officer*, 1909: 11.

[51] *Report of the Veterinary Officer*, 1909: 8, 11.

[52] *Report of the Veterinary Officer*, 1909: 7–8 and 1910: 1, 3.

[53] *Report of the Veterinary Officer*, 1909: 12.

[54] *Annual Administrative Report of the Civil Veterinary Department for the Official Year 1910–1911* (Calcutta: Superintendent of Government Printing, India, 1911), 5.

[55] For a history of the establishment of the laboratory and its early activities, see J. E. D. Holmes, *A Description of the Imperial Bacteriological Laboratory, Muktesar: its Work and Products* (Calcutta: Superintendent of Government Printing, India, 1913).

[56] According to Benjamin Moore, the German firm was Lanolinfabrik; see "Atoxyl and Soamin and the Treatment of Sleeping Sickness and Syphilis" , *British Medical Journal* 2510 (Feb. 1909), 370–371. As near as I have been able to ascertain, one of the earliest reports on the use of Atoxyl appeared in Great Britain was in 1907; see Anton Breinl and John Todd, "Atoxyl in the Treatment of Trypanosomiasis" , *British Medical Journal* 2403 (Jan. 1907): 132–134.

[57] On Leese's initial experiments, see A. Leese, "Summary of First Series of Experiments on the Treatment of Surra in Camels" , JTVS 5 (1910): 57–64 and 397–410, which involved alternating Tartar Emetic with Atoxyl and Mercury Bichloride; see *Report of the Veterinary Officer*, 1911: 4–5.

[58] *Report of the Veterinary Officer*, 1912: 2–9, which includes a list of instructions for administering the drugs; and A. Leese, "Third Series of Experiments on the Treatment of Surra in Camels, with Some Cures" , JTVS 7. 1 (1912): 1–18. On surra research at Muktesar, see J. D. E Holmes, "The Cure of Surra in Horses by the Administration of Arsenic" , JVTS 6. 4 (1911): 447–467.

[59] *Report of the Veterinary Officer*, 1911: 5 and A. Leese, "Third Series" , 8.

[60] *Annual Report of the Camel Specialist for the year 1912–1913* (Lahore: Punjab Government Press, 1913), 2. Hereafter cited as *AR Camel Specialist* with year of publication.

[61] *Annual report of the Punjab Veterinary College, Civil Veterinary Department, Punjab and the Government Cattle Farm, Hissar for the year 1914–1915* (Lahore: Superintendent Government Printing, 1915), 7.

[62] On rakh policies, see Gilmartin, *Blood and Water*, 81, 157–158.

[63] *Annual Reports for the Chenab, Jhang, Chunian and Jhelum Colonies, 1908* (Lahore: "Civil and Military Gazette" Press, 1909), 1.

[64] *Annual Reports for the Chenab, Jhang, Chunian and Jhelum Colonies, 1910* (Lahore: "Civil and Military Gazette" Press, 1911), 1.

[65] *Annual Reports for the Chenab, Jhang, Chunian and Jhelum Colonies, 1914* (Lahore: Punjab Government Press, 1915), 2.

[66] IOR/P/8672: July 1911-December 1911. "Report of the Camel Browsing Committee," *Punjab Proceedings, Department of Revenue and Agriculture* Programs, December 1911, no. 13, 6. Also see Imran Ali, *The Punjab Under Imperialism, 1885–1947* (Princeton: Princeton University Press, 1988), 129.

[67] Maj. F. S. H. Baldrey, "The Evolution of *Trypanosoma evansi* through the Fly" , *Journal of Tropical Veterinary Science* 6. 3 (1911): 271–282.

[68] Fellow Royal Institute of Public Health, Fellow Royal Civil Veterinary Society, and Fellow Royal Geographic Society. He had also been involved in producing the 9th edition of William Williams' *The Principles and Practices of Veterinary Medicine* (1909).

[69] *Annual Report of the Veterinary Officer*, 1911: 2–3. 1908 and 1909 were La Niña years, which meant increased water volume in the monsoon that swept across the Punjab.

[70] *Sleeping Sickness Bulletin* 3. 30 (1911): 367–368.

[71] A. Leese, "Biting Flies and Surra" , JTVS 7. 1 (1912): 19. I cite page numbers from this source in the text.

[72] A. Leese, " 'Tips' on Camels, for Veterinary Surgeons on Active Duty" , *Veterinary Journal* 73 (1917): 86.

[73] See, for example, Hoare, *Trypanosomes*, 558 and Lorne E. Stephen, *Trypanosomiasis: A Veterinary Perspective* (Oxford: Pergamon Press, 1986), 187.

[74] Leese, *Out of Step*, 18.

[75] I take the quotation from Warwick Andersen, where he, of course, is speaking of humans under the care of the American colonial regime in the Philippines. *Colonial Pathologies: American Tropical Medicine, Race and Hygiene in the*

Philippines (Durham: Duke University Press, 2006), 6.

[76] J. E. D. Holmes, "Investigation of an Outbreak of Horse Surra with Results of Treatment with Atoxyl, Tartar Emetic, Mercury and Other Drugs" , JTVS 3 (1908): 158–172.

[77] *AR Camel Specialist*, 1913: 4.

[78] *AR Camel Specialist*, 1914: 11–12.

[79] See IOR/V/25/125: *Bulletin no. 95, Agriculture Research Institute, Pusa*, "A Note on the Treatment of Surra in Camels by Intravenous Injections of Tartar Emetic" , (1920): 1–4. He also carried on or directed numerous studies of the habits of biting flies, only to conclude that surra was, in fact, transmitted mechanically. See IOR/V/25/541/31–35: H. E. Cross and G. Patel, "Surra Transmission Experiments" , *Department of Agriculture, Punjab. Veterinary Bulletin No. 5 of 1921*: 1–13; and H. E. Cross, "A note on the Transmission of Surra by Ticks," *Department of Agriculture, Punjab. Veterinary Bulletin No. 6 of 1921*: 1–3; "A Note on the transmission of Surra by Tabanus nemocallosus," *Department of Agriculture, Punjab. Veterinary Bulletin No. 7 of 1921*: 1–7; and "Surra Transmission experiments with Tabanus Albimedius and Ticks" , *Department of Agriculture, Punjab. Veterinary Bulletin No. 12 of 1923*: 1–11.

[80] One also ought not to discount the arrival of the Nobel Prize in Physiology or Medicine, or other prizes, as factors in rivalries. Four of the first eight prizes in this category, beginning with Ronald Ross, who received his prize for establishing the relation between the malarial parasite and anopheles mosquito, went to researchers either directly or indirectly connected to tropical medicine research. In addition to Ross, they were Robert Koch, who visited Muktesar in 1898 and was involved in trypanosomiasis research in Germany's African colonies, Alphonse Laveran, author of the first comprehensive work on trypanosomes, and Paul Erlich, who received his prize for work on immunology.

[81] Michael Worboys, "Mason, Ross, and colonial medical policy: tropical medicine in London and Liverpool, 1899–1914" , 21–37, in Ray Macleod and Milton Lewis, eds. *Disease, Medicine and Empire: Perspectives on Western Medicine and the Experience of European Expansion* (London and new York: Routledge, 1988), 22.

[82] John Henry Steel provided tips for veterinarians on locally available drugs and herbals; see, for example, "Materia Medica Veterinaria Indica" , *Quarterly Journal of Veterinary Science in India* 4. 16 (Jan. 1886): 415–424.

[83] Thomas Pynchon, *Gravity's Rainbow* (London: Vintage Books, 1973] 2013), 296.

[84] Atoxyl was administered to Africans suffering from Sleeping Sickness and confined to "segregation camps" . Some went blind. See the discussion of the scholarship on the subject in Deborah Neill, *Networks in Tropical Medicine* (Stanford: Stanford University Press, 2012), 110-116.

[85] Testing of Naganol began by 1926 in India. See J. T. Edwards, "The Chemotherapy of Surra in Horses and Cattle in India, *Journal of Comparative Pathology and Therapeutics* 39.2 (June 1926): 82-112 and 39.3 (Sept. 1926): 169-201. For accounts of successful treatments with Naganol, also see S. K. Sen, "Prophylaxis against Equine Surra by Means of 'Bayer 205' (Naganol)" , *Indian Journal of Veterinary Science and Animal Husbandry* 1. 4 (1931): 283-295; hereafter, IJVSAH.

[86] Quartermaster General's Branch, Army Headquarters, India, *Handbook on Contagious and Infectious Diseases of Animals* (Calcutta: Government of India Central Publication Branch, 1929), 297-303.

[87] *Punjab Colony Manual* (Lahore: Superintendent of Government Printing, Punjab, 1922), v. 1: 52-55.

[88] See IOR/L/MIL/6780, which is labeled "Reorganization of Animal Transport, 1924-1930" . The key document for the numbers given here is dated 22 May 1930.

[89] Quartermaster General's Branch, Army Headquarters, India, *Hand Book on Contagious and Infectious Diseases of Animals* (Delhi: Manager of Publications, 1936), 296, 298.

[90] W. Taylor, "Surra in the Punjab" , IJVSAH 4.1 (1934): 29.

[91] Bachan Singh, "Bovine Trypanosomiasis in Central Provinces with An Account of Some Recent Outbreaks" , IJVSAH 6.3 (1936): 243-250.

[92] Robert Jensen, Larry Simpson and Paul Englund, "What happens when *Trypanosoma brucei* leaves Africa" , *Trends in Parasitology* 24.10 (Oct. 2008): 430.

[93] Muhammad Ali A Shah, Kalili Ur Rehman, Fawad Ur Rehman, and Nongye He, "Present Status of Camel *Trypanosomiasis* in Pakistan, a review of literature" , *Science Letters* 1.1 (2013): 30-33.

[94] S. Ahamad, et al., "The Economic importance of Camel: A Unique Alternative under Crisis" , *Pakistan Veterinary Journal* 30.4 (2101): 191-197 and Khan, Bakt Baidar, *Production and Management of Camels* (Faisalbad: University of Agriculture, 203).

[95] The connection between increases in malarial infections and canal expansion had been made as early as 1911; see S. R. Christophers, *Malaria in the Punjab* (Calcutta: Superintendent of Government Printing, 1911): 101-104.

学人访谈

Judith Farquhar Interview: Reflections on Research, Writings, and Medical Anthropology

Zhang Wenyi

Editor's Note: Judith Farquhar is Max Palevsky Professor Emerita of Anthropology and of Social Sciences in the Department of Anthropology at the University of Chicago. She does research on traditional medicine, popular culture, and everyday life in contemporary China. Anthropological areas of interest include medical anthropology; the anthropology of knowledge and of embodiment; science and technology studies; critical theory and cultural studies; and theories and practices of reading, writing, and translation. She is the author of *Knowing Practice: The Clinical Encounter of Chinese Medicine* (1994)《认识实践：遭遇中医临床》, *Appetites: Food and Sex in Post-Socialist China* (2002)《饕餮之欲——当代中国的食与色》（江苏人民出版社，2009，中文版）, and *Ten Thousand Things: Nurturing Life in Contemporary Beijing* (co-authored with Zhang Qicheng, 张其成 2012)《万物·生命：当代北京的养生》

（三联书店，2019，中文版）. She co-edited *Beyond the Body Proper: Reading the Anthropology of Material Life*《超越肉体：物质生活的人类学读本》as well as several journal special issues. She is currently working with Lai Lili on a forthcoming book *Gathering Medicines: Nation and Knowledge in China's Mountain South.* Professor Zhang Wenyi at Sun Yat-sen University conducted an interview with Professor Judith Farquhar in 2018, The following is the interview transcript in its entirety.

Thank you, Zhang Wenyi, for posing such wide-ranging and challenging questions. You have paid close attention to my work, and I'm very flattered that you think it is important. In a sense, your questions presume a familiarity with my published work; if I were to answer adequately, for all readers, I would have to write rather lengthy essays on each point of each reply. Since that is not feasible, I will just offer a few slogan-like position statements and contextualizing comments, and hope that our readers can guess the wider arguments.

Zhang: Prof. Farquhar, thanks for accepting an interview concerning your research in anthropology and Chinese medicines. I would like to begin our conversation with a banal question : how did you become an anthropologist, and how do you live as an anthropologist, in addition to doing fieldwork and writing ethnography?

Farquhar: I could tell many different stories about how I became an anthropologist, but even I am not sure which of them is really a true history.

I remember reading some ethnography when I was 15, and deciding that this kind of investigation could begin to satisfy my curiosity about the world beyond my rural community. I also had a chance to appreciate Chinese painting and Chinese literature in translation while I was still in high school. So perhaps I had an "anthropological" point of view from an early age, and during my short time in college I majored in anthropology. But I really "became an anthropologist" only after I had been working for 10 years at the National Institutes of Health, and my supervisor there — who was himself a researcher who traveled to remote parts of Oceania — urged me to return to university. When I started reading for my anthropology courses in graduate school, I realized that anthropology offered a blend of abstract principles (or "theory") and concrete stories; this seemed right, or true, to me. The field and its methods, and its writing style, "fit" with a way of seeing and appreciating things from all over the world that I had been learning for a long time.

I often tell students who are not sure whether to commit themselves to graduate work in anthropology that the life of an academic anthropologist is pretty close to perfect. A university or college pays us to read, write, and teach — all of these activities are enjoyable opportunities to learn and grow. And we can get research grants to travel to the parts of the world we find most fascinating, talk to all kinds of people about their lives and their thoughts, share food and labor and living spaces, and dig for something true beneath the superficial over-simplifications we get from the mass media. It takes a long time to become an academic anthropologist, and university jobs are too few (though academic anthropology in China is growing fast and becoming a good place to practice our craft). But the investment in a life of indulging curiosity and seeking knowledge in depth is worth it.

Zhang: So nice to hear your joy of being an anthropologist. Now, let's turn to the source of your joy — your professional practice in anthropology. I have noticed that in your works from *Knowing Practice*, through *Appetites* and *Ten Thousand Things*, to your forthcoming *Gathering Medicines*, you (together with your co-authors) have been working to understand social practices through "the body" by embedding it into social transformations and by returning back to the materiality of "the body." Could you please tell us some turning points in your career that led you to focus on the body and embodiment?

Farquhar: When I was in graduate school at the University of Chicago, my teacher Jean Comaroff suggested that I think about presenting my work on Chinese medicine as an anthropology of the body. At the time, I had no idea how this would be done. And my first book was not really about "the body" , because it was really analyzing knowledge as a practical process, one that was discernible in my observations of (and literary study of) clinical work.

Even so, I remember, in the course of a train trip from Guangzhou to somewhere in the north in 1983, looking out the window in the early morning before everyone else in the hard berths was awake, thinking about the classical Cartesian "body/mind" divide, and deciding — "there is no 'body' in China" . This ridiculous idea was my own rejection of Cartesianism for Chinese medicine, but it was also — paradoxically — a beginning of thinking in a carnal or materialist way about lived embodiment in China. Eventually I developed a way of thinking about bodies as concentrations of practices which extend beyond the skin-boundary and put the whole body — inward turning and outward-reaching — into motion, treating it as a condensation of life rather than as a structure. So I began to talk about embodiment or bodiliness, not "the body" , trying to emphasize the active contingency and specificity of all manner of bodies.

It is regrettable that everyone in anthropology seems to think that an anthropology of medicine is necessarily an "anthropology of *the body*" . Classic African ethnography showed us long ago that healing is about many kinds of affliction. Gods and ghosts, crimes and cruelty, symbol and ritual are as important as symptoms and drugs in the anthropology of medicine. Therefore, if we are going to ask medical anthropology to be "about bodies" , we must think of bodies as formations of lived social practice.

Zhang: I would like to rephrase your point, if I understand you correctly, by saying that one theoretical question that weaves through your diverse research projects is whether and how the body is local (there are bodies of peoples), rather than exclusively universal (there is the body of human beings). Could you please give us an example from your fieldwork illustrating that the body is local? And furthermore, how does your approach differ from Margaret Lock's concept of "local biologies" , if in particular, we consider that you and Prof. Lock co-edited the volume entitled *Beyond the Body Proper* proposing to go back to the materiality of the body?

Farquhar: Margaret Lock's notion of local biologies has been a very important way of relativizing embodiment. She was able to show, comparing women's mid-life symptoms and concerns in Japan and North America, that carnal experience (and, of course, disease) could vary a great deal in these contrasted cultural worlds. Our medicalized social science needed these reminders so researchers could move beyond the modern common-sense assumption of a universal human body, structured and lived — and subject to pathology — in the same way everywhere.

But the term "biologies" might mislead just a little. Let me give you an example of how a "practice body" is always and everywhere "local" . We don't have to go far. Thinking about your question, I was riding the subway in Beijing and noticing again how many unique cyborg networks are gathered in a place like that. Everybody (every body) was using their smart phone to keep afloat in a weblike network of connections. The meaning of these lives, even their social health, was completely contingent on a vast prosthesis (a system of artificial body parts) called the phone and the internet. Think about how disoriented and cut off (amputated?) you feel when you cannot get online. Your WeChat network and mine overlap, but as unbounded wholes they are unique. And these interpenetrating particularities are essential to our sense of our particular life. So the practices that make us "local" as bodies, as a thousand webs of connections anchored in a thousand human beings, are certainly concrete and material, but they cannot be seen as a universally "same" and discrete structure like "the body" . Biology is not irrelevant, but it is not the only science that knows "the body" . If we're really interested in relativizing embodiment, after all, biomedicine's biology is not the only systematic and powerful conceptual language available.

Zhang: I love your interpretation of everybody as every body, which beautifully illustrates your idea of local bodies. To explore diverse bodies and embodiment, you gleaned inspirations from Chinese medicine and Chinese philosophy, like that of Zhuangzi, both in your *Ten Thousand Things* and *Gathering Medicines*. Please give us one or two examples illustrating how clues from Chinese philosophy facilitate you to construct an interesting and novel understanding/analysis on bodies-in-practice otherwise impossible if we follow anthropologists' conceptualization of the body inherited from Mauss, Foucault and Bourdieu?

Farquhar: Following on my thoughts above about locality and biology, we can ask whether the European authorities you mention are still indebted to a Cartesian body/mind ontological divide. I don't think it is impossible to expand the notion of embodiment in company with Mauss, Foucault, and Bourdieu. I have read all three of these inspiring teachers somewhat against the grain of the English and French languages to find an irreducibly social (and thus networked) body, always relational, always under construction, always subject to dispersion or disruption. I feel that they are anti-Cartesian, each in their own way. But they have fought an uphill battle against the dualist ontology of modernist European languages.

When you turn to Chinese sources, it is much easier to speak of the relational (and dynamic, and lived) body, clearly and without contradictions. In my work I have always taken conceptual inspiration from the language of Chinese medicine, which can speak of so much more than "biologies" or "the body." In *Knowing Practice* I insisted on centering local Chinese notions of practice and experience as I constructed the focus of my ethnography; These concepts differed somewhat from those of modern practice theory (Bourdieu) or individualist phenomenology (e.g. as discussed by Byron Good). In *Appetites* I devoted a chapter to the Chinese medical dyad of excess and deficiency (虚 and 实), showing how one can quantify in this way even when process and flow are the only reality for either healing or social life (the chapter also drew on some stories of excess and deficiency explored by author Mo Yan). And in *Ten Thousand Things* Zhang Qicheng and I emphasized the cosmic generativity of classic philosophy, showing how "life giving birth to life, transformations upon transformations" (生生化化) was an ancient principle that helped to make sense of the wellness activities of Beijingers. Now, working with Lili Lai on *Gathering Medicine*, we are tracing patterns of *gathering* and *scattering* in the emergence of nationality medicines: in this we are following Zhuangzi, though Martin Heidegger and Bruno Latour are not far off.

Zhang: Let me take the idea of networked bodies a little farther back by contextualizing it into, or against, the conceptual development of anthropology. One of the reasons for anthropologists' interest in practice and the body is to go beyond the enduring divisions in anthropological thought regarding nature/culture, the social/biological, and system/subject. In your upcoming book on herbal medicines in China, you propose to understand herbs and their efficacy as an intimacy cultivated in an assemblage of herbs, persons, and the activities of collection and processing herbs. In what sense do such intimacy and gathering imply something that fuses some theoretical divisions in anthropology?

Farquhar: I really appreciate your understanding of our project, your summary makes me feel that we might be able to see this book as a contribution to anthropology at large. Of course I feel that it has been part of our task in anthropology to "fuse some theoretical divisions" while continuing to make clear and useful distinctions in a way that is materialist and empirical. For me, the great looming features of what Latour has called the modern constitution that must be (at least) reconsidered are the following dyads: individual and society, body and mind, and (following William James the pragmatist as early as 1907) thing and thought. Latour argues that "we have never been modern" in the sense that these modernist categories have never been quite right for describing our everyday realities. The Chinese pre-modern constitution may be equally off-the-mark with its big "Confucian" categories (consider the beautiful but possibly unrealistic ideals of the *Great Learning*《大学》for example), but at least it has not misled us into naturalizing "the individual", "society", "body", "mind", and the rest. By trying, with our ethnography of the development of nationality medicines in China's mountain south, to appreciatively read and cite the practice and the insights of local healers, perhaps we are in search of a different conceptual constitution for the real lives of numerous actors, human and non-human alike.

Zhang: You used the word "appreciatively" , which, it seems to me, suggests a peculiar relation between you, as a researcher, and the people you are studying. In what sense could an appreciative relation be maintained and balanced with your materially and empirically-based and theoretically-driven ethnographic research?

Farquhar: Well, thank you for noticing! One of the anthropological book chapters I often teach is Talal Asad's "The Concept of Cultural Translation in British Social Anthropology" , which appears in George Marcus and James Clifford's influential book, *Writing Culture*. I could justify my commitment to an appreciative stance on "the people I am studying" with reference to Asad, who in this article denounces the condescending critique characteristic of some British social anthropologists.

One could do a long history of the negative critical stance intrinsic to anthropology's fetishization of "the primitive" and even "the oriental" . (But others have done that, notably Edward Said and Johannes Fabian.) We are very familiar with an anthropology in which "the West" consolidates its cultural superiority through a kind of subtle denunciation of different life ways and systems of ideas.

The alternative to such an Orientalist anthropology is not easy to work out as a methodology. But one starting point might be to aim at simple appreciation of difference. To appreciate is to see, to value, even to esteem. It is a willingness to *let be*: even as the outsider anthropologist avoids any temptation to romanticize "the native" , which is just another side of Orientalism, she and her interlocutors can experiment with coexistence in plural worlds and cooperation in knowledge production.

Even for the "native anthropologist" , the fieldworker who hails from an elite urban university in China and "studies" migrant youth, or village economies, or ethnic medicine in the P. R. C., there may be deep cultural divides that need to be crossed somehow. Cultural translation is never really frictionless. But I have long felt that it is not our job — we who belong to the privileged elite who are allowed to speak about society — to *denounce* a local world in which we have no long-term stake. I return to Chicago, you return to Guangzhou, both of us actually live in universities. Critique is incumbent upon those inside the system to which they are non-optionally subject.

Zhang: Medical anthropology has been one of the most developed while continuously increasing subfields of anthropology. I am also working in medical anthropology, and often wondering how medical anthropology could contribute to anthropology at large, which medical anthropologists could most clearly demonstrate in the domain of health and illness, and which could stimuate anthropologists in other domains to think about their materials in a novel way. To use a not-so-appropriate phrase, does medical anthropology have some essence that distinguishes it from other subfields of anthropology?

Farquhar: As you know, it is easy to replace "essences" proposed by analysts with something more historical and empirical: we need not resort to something like "human suffering" as a core problematic of medical anthropology, for example, because "medicine" itself is a much more sturdy historical domain demanding our attention. What I mean is, medicine as a biological science and form of clinical expertise is relatively young in European and American history (see Foucault, *The Birth of the Clinic*). Also, the debates about the meaning and scope of "yi 医" in China are old, but they are, precisely, *debates* and reflections on whether there even is a systematic or bounded discursive field called "medicine".

The domains of practice we can properly call medical today have — for the time being — special features that all anthropologists would do well to pay attention to: modern medicine, for example, has a strong tendency to individualize experience and ground it in an evolving view of anatomy. Both traditional medicines and the social sciences have learned this view of things from the history of dominant biomedicine. But since the 1980s, at least, medical anthropology has been interrogating the naturalized categories of the modernist social sciences, refusing entities such as the autonomous individual and the structured/mechanical body. Our findings should make it more unlikely for functionalist anthropologists to "explain" social activity with reference to "biological needs". And perhaps more than other specialties, we medical anthropologists have shown that there is nothing universal — not even "the body" — that could be called human nature.

Still, because our field is that of affliction and healing, and because we engage both expert knowledge and popular "common sense", we must address many difficult questions of theory and method. Our problems thus overlap with those of bioethicists, policy-makers, social workers, and even medical doctors themselves. Other anthropologists are not our only public.

Zhang: Now, we turn to your experience as an anthropologist in this global era. In what sense does your experience in China influence your life in the States as an anthropologist, as a university professor, and as a citizen? Do you separate your life as a citizen in the United States and as an anthropologist doing fieldwork in China? If so, how?

Farquhar: This question is hard to answer, especially since I am going back and forth between Chicago and Beijing so much lately. To the extent that I experience my life as a continuity, these days my "two lives" don't seem very different. At first, perhaps, in the 1980s and 1990s, while I was still mainly involved with Chinese medical people, I didn't feel much like an academic anthropologist while I was in China. My friends and associates in medical colleges and hospitals often had complicated intellectual lives, but few of them were cosmopolitan "social scientists" . They would not have understood what I was writing in English about their practice and experience. So, working with experts in the field of Chinese medicine, and — for a while — with local leaders and doctors in the Shandong countryside, I could sometimes feel like a philosopher, or like a clinician, but anthropological concerns (such as "theory" and pedagogy) were usually far from my mind. In the last decade or so, though, I have been quite engaged with anthropologists and social theorists in Beijing and elsewhere in China, so life in my two cities has converged for me.

But you ask about fieldwork: because Lili Lai and I have been doing fieldwork in the south of China, where there are many local dialects that I cannot follow, and because we have never stayed in one place for long enough to become really expert in local lifeways, I have often felt quite the stranger in some of the places we have visited. Even so, some of our field contacts have very kindly welcomed us into their lives. In those cases, I have begun to feel like a student of very particular kinds of expertise, and I have developed a deep respect both for herbalists and for local researchers. Is that what it is to feel like an anthropologist?

Zhang: I think so. And I think maybe there is also another way of feeling like an anthropologist when we come back from fieldwork. We have to tell other anthropologists, or even scholars in other disciplines, that our specific ethnographic case study is meaningful in other contexts. Could you share with us some of your cases that move from ethnography to anthropology?

Farquhar: A few readers of my book *Appetites* have read it as a demonstration of a certain method in anthropology. They have told me that they can see, after reading it, how one might understand bodies in history by attending to all kinds of down to earth, everyday actions and by reading a diversity of mediated forms (fiction, films, advertising, propaganda, public education...) to inquire into the social-material conditions that make such actions and forms possible. This assessment of my book has meant a great deal to me. In planning the book I had decided to look at food, sex, and — always! — medicine as especially privileged windows on to carnal-practical life. (There were originally to be ten different topics including architecture, history textbooks, and much more, with food and sex being only two chapters. But my friend John McGowan persuaded me that food and sex would keep me occupied for the length of a book, as they certainly did.) Anthropologists often speak of the body as silent in history; the scholar needs to trick the archive into yielding concrete information about how life is lived in a particular time and place. I guess I saw *Appetites* as a demonstration project in tricking the surfaces of mediated social life into showing deeper aspects of material embodied life. So I especially appreciated my readers who saw the book as methodology. Given that I reject the anthropology of a universal human nature practiced by those who follow the example of Levi-Strauss, this contribution to method might be as close as I can get to anthropology as the study of Man (committed as I am to China and specificity)!

Zhang: Readers experience inspiring moments in reading your books, that's great. And I would like to hear about some inspiring moments brought by anthropology in your life that encouraged, excited and vitalized you, your colleagues and your students. As we often heard, A discipline that does not keep stimulating the imaginations of younger generations will be doomed.

Farquhar: Much of my inspiration comes from things that happen in teaching. There has been no shortage of thrilling moments for me, when I have witnessed students grappling with ideas and descriptions that completely take them by surprise. One of the earliest such moments was when I was teaching my first "core" social theory class to first year students in the College at the University of Chicago. After a few weeks of culture theory and a little semiotics, one of the students stopped me after class, as I was erasing the blackboard. He said: "I'm confused! I want to get this straight — it seems to me that for you anthropologists, everything counts. Is that right? Does even the most trivial thing have meaning in this anthropology game?" I could only laugh and say "Yes, I guess that's right." I have often thought back to that moment as a sort of turning point. After that I could not settle for a reductionist social science that would abstract thinking or writing away from the concrete details of life. For us, everything, no matter how superficial, *counts*. I am thankful for that student's frustration and insight.

Zhang: One last question, what is the importance in your work of collaboration and co-authorship with writers based in China and writing in Chinese?

Farquhar: I cannot overstate the importance of collaboration for anthropology. My work with Zhang Qicheng on the field work and the planning of *Ten Thousand Things: Nurturing Life in Contemporary Beijing*, was immensely rewarding, and the process taught me a lot. When, after I had written the English language manuscript, Prof. Zhang edited and then warmly approved every paragraph, I felt uniquely validated as an anthropologist and a thinker. Then he approved a careful Chinese translation; we published this with Sanlian Press in the middle of 2018. Because of Prof. Zhang's participation, this book in both languages is a contribution to much more than academic anthropology, we feel.

As for my current collaboration with Lai Lili, the book itself will demonstrate how central our cooperative relationship has been. I hope readers will not expect confessions of quarrels and conflicts, though; in fact we have been mostly in accord — in the field, in outlining the book (again and again!), and in designing and drafting the chapters. But Lili reads with great critical acumen; she catches every bit of sloppy thinking in the draft and insists on starting over when I have written my way into blind alleys. Conversations with Lili send me to the keyboard with excitement and a sense of adventure; her field notes inspire me to write more accurately; and her erudition thickens our "field" in wonderful ways.

Every anthropologist should have a co-author located closer to the sites of their research. I no longer fully trust ethnographies written by solitary foreign experts.

理论与实践

生态博物馆与博物馆人类学：回溯与反思

尹绍亭*

我国最早的生态博物馆诞生于贵州，它的创始人是苏东海和胡朝相先生。贵州生态博物馆的建设与笔者倡导的文化生态村建设几乎同时出现。20 世纪 90 年代后半期，以社区、村落文化生态保护为宗旨的贵州生态博物馆和云南民族文化生态村同时出现于中国西南邻近的两个省份，并非偶然。它与两省得天独厚的自然生态环境、丰富多彩的民族文化、文化生态研究深厚的积累、开放包容的国际学术交流氛围等条件是分不开的。苏东海是资深博物馆专家，胡朝相是文物管理研究专家，与我是同行，都对博物馆事业情有独钟。此外，我又是研究生态人类学的，对“生态”二字特别敏感，所以当苏东海第一次向国人介绍“生态博物馆”时，我便被深深吸引。比较“贵州生态博物馆”和“云南民族文化村”两个文化生态保护模式，既有共同的方面，也有诸多差异。美美与共，各美其美；他山之石，可以攻玉。在我们建设云南民族文化生态村的过程中，曾不断比较参考生态博物馆，借

* 尹绍亭，云南大学民族与社会学院。

鉴其经验教训，相互之间交流不断。弹指之间，这两项事业不觉已经过去了20余年，现在虽然我们基本上已不在一线工作了，但是作为曾经倾注过大量心血的事业，总是难以忘怀。时间是检验器，许多心得和反思是靠时间打磨出来的。本文再次回顾贵州生态博物馆，除了评述之外，还试图从博物馆人类学的角度做一些探讨。

一、苏东海和生态博物馆

苏东海先生是著名博物馆学专家，曾任原中国革命博物馆陈列部主任、《中国博物馆》、《中国博物馆通讯》杂志主编等，长期致力于博物馆学和博物馆发展研究，主要著作有《博物馆演变史纲》《博物馆》《中国博物馆哲学》《博物馆的沉思——苏东海论文选》《贵州国际生态博物馆论坛论文集》《苏东海论文选登》等。苏东海在国内外被称为"中国生态博物馆之父"。

1986年，苏东海首次在《中国博物馆》杂志上发表生态博物馆的文章，将发端于欧洲的这一新型的博物馆模式介绍给学界。时任贵州省文物处处长的胡朝相早就有建设社区博物馆的想法，获得生态博物馆的信息，便积极联系请教苏东海。1995年，在苏东海的悉心指导帮助协调下，通过胡朝相的努力，贵州省生态博物馆建设提上省政府的议事日程，并得到挪威政府的援助，被纳入"1995至1996年挪中文化交流项目"得以实施。贵州生态博物馆建设选择了四个地点：梭戛［苗族］、镇山［布依族］、隆里［汉族］和堂安［侗族］。贵州生态博物馆在苏东海的推动下开局顺利，形成了很好的势头。

贵州生态博物馆起点甚高，为国家层面的国际合作项目——中国挪威文化合作项目。1997年10月23日，中国国家主席江泽民和挪威国王哈拉尔五世、王后宋雅在北京人民大会堂出席了《挪威开发合作署与中国博物馆学会关于中国贵州梭戛生态博物馆的协议》签字仪式。中国国家文物局张文彬局长和挪威外交大臣沃勒拜克分别代表中国和挪威政府在协议上签字。出席签字仪式的还有国务院副总理钱其琛、全国人大副委员长王汉斌、全国政协副主席朱光亚等。一个“文化协议签字仪式”，有如此之多的中挪国家领导人参加，尤其是江泽民国家主席和挪威国王皆亲临会场，非同寻常。

贵州生态博物馆项目由苏东海担任项目领导小组组长，项目组成员有博物馆专家安来顺、文物专家胡朝相等，国家文物局马自树副局长担任项目领导小组顾问，贵州省政府副秘书长曹新中担任项目实施小组顾问，项目成员还包括生态博物馆所在地各级政府官员。项目自始至终得到挪威专家的参与指导，曾在挪威担任过15年生态博物馆馆长、在欧美指导帮助过若干生态博物馆建设的著名博物馆学家约翰·杰斯特龙先生受聘为贵州生态博物馆科学顾问。贵州生态博物馆项目组中外学者结合、官员学者兼备，实属难得。经费保障是贵州生态博物馆的又一个优势。别的不说，仅梭戛苗族生态博物馆的前期建设经费，就有挪威政府和国内支持的286万元。1999年3月，挪威开发署又决定再次提供150万挪威克朗，以资助建立贵阳市花溪区镇山村、锦平县隆里古城、黎平县肇兴乡堂安生态博物馆的规划和帮助梭戛生态博物馆的进一步开发。2001年10月，挪威政府再次决定对隆里、堂安两座生态博物馆无偿援助300万元人民币，贵州省政府亦决定配套500万元人民币（胡朝相，2011：38）。2005年我们前往梭戛参观，看到新建的大片崭新民居以及村口的接待站等，印象十分深刻，这些工程没有数百万乃至上千万资金的投入是不可能建成的。

贵州生态博物馆是由苏东海倡导、筹划、促成的，可以说没有他就没有贵州的生态博物馆。此事很能说明专家学者的特殊作用。顺便说一句。苏东海早年曾在昆明读书，也算是从云南走出去的学者，与云南有不解之缘。如果云南早于贵州尊重他、请教他，此项目很可能就会落户于云南而不是贵州。生态博物馆和云南失之交臂，至今思来殊为遗憾。

二、什么是生态博物馆

从1989年至2014年，因为从事云南民族博物馆和云南民族文化生态村的建设，笔者曾经考察过日本的近百个城市和社区的博物馆、乡土馆，法国的数十座博物馆以及澳大利亚、加拿大、韩国等国的许多博物馆。也许是由于生态博物馆在这些国家尚不是主流，所以相关信息所获不多。关于它的资讯，主要是参考苏东海、胡朝相等人关于生态博物馆的若干论述和学界的讨论，兹概括简介如下。

讨论生态博物馆，首先需了解其产生的背景。20世纪60、70年代，工业社会在经历了辉煌之后，其对社会思想、文化遗产、生态环境和自然资源等的消极影响日益显现，社会性的危机感、焦躁感悄然涌动，以致形成了一股强大的波及社会各界的反思和批判的潮流。社会层面，以美国生态学家雷切尔·卡尔逊的《寂静的春天》出版为标志，公众环境保护意识觉醒，当代环保运动随之兴起。与此同时，思想界、学术界也把目光投向了生态环境保护，人与环境、人与生物圈的研究蓬勃发展。生态文明建设、政治生态学、社会生态学、环境人类学、环境史等新研究领域应运而生。在这样的背景下，博物馆作为面向公众教育和科普的文化机构必然会做出反应，生态博物馆便是博物馆界适应新时期生态文化需求的一个新的创造。

生态博物馆于20世纪60年代最早产生于法国。被称为生态博物馆之父的法国博物馆专家乔治·亨利·里维埃（Georges Henri Riviere）是这样定义生态博物馆的："通过探究地域社会人们的生活及其自然环境、社会环境的发展演变过程，进行自然遗产和文化遗产的就地保存、培育、展示，从而有助于地域社会的发展，生态博物馆便是以此为目的而建设的博物馆。”另一位法国博物馆学家雨果·黛瓦兰（Hugues de Varine）认为：“生态博物馆是居民参加社区发展计划的一种工具。”法国的《生态博物馆宪章》把生态博物馆定义为：“生态博物馆是在一定的地域，由住民参加，把表示在该地域继承的环境和生活方式的自然和文化遗产作为整体，以持久的方法，保障研究、保存、展示、利用功能的文化机构。”对于生态博物馆的进一步的解释，见于乔治·亨利·里维埃所概括总结的《生态博物馆的发展的定义》，他在该文中写道：生态博物馆是行政当局和住民共同构想、创造、利用的手段（尹绍亭，2008）。

从上面的介绍可知，相对于西方传统博物馆的贵族性、殖民性、都市性、国家性、垄断性等特征，生态博物馆有许多可以称之为“革命性”的突破。对于两者理念体系的差异，约翰·杰斯特龙曾做过如下简练的对比：生态博物馆的保护研究展示对象是自然文化遗产，传统博物馆的保护研究展示对象是藏品；生态博物馆的运作空间是社区，传统博物馆的运作空间是博物馆建筑；生态博物馆的服务对象是社区住民，传统博物馆的服务对象是观众；生态博物馆注重文化记忆和公众知识，传统博物馆注重科学知识及科学研究。

生态博物馆在法国产生，创造了不同的类型，形成了较为完整的理论、方法和管理体系，并在世界上很多地区产生了影响。1980 年以后，生态博物馆为法语圈、西班牙语圈、葡萄牙语圈、意大利亚语圈以及拉丁语系的许多国家所接受，其理念在欧洲、北美洲、南美洲、非洲、大洋洲和亚洲得到了普及，出现了迅速发展的势态，迄止 20 世纪 90 年代，全球的生态博物馆数量曾一度达到 300 多座（苏东海，1999）。

三、贵州生态博物馆及其影响

早在1995年贵州生态博物馆建设之初，笔者便注意收集与其相关的新闻报导和资讯，并查阅了《中国博物馆》、《中国博物馆通讯》等杂志所刊登的所有相关专题文章。梭戛苗族生态博物馆于1996年开始筹建，曾多次想去实地考察，苦于一直没有机会。直到2005年6月应邀参加贵州省举办的“贵州生态博物馆国际论坛”，才去参观了梭嘎和镇山的生态博物馆。2010年夏天，又承蒙胡朝相先生邀请，在他的陪同下，先后考察了隆里古城生态博物馆、堂安侗族生态博物馆以及民间主导的地扪侗族生态博物馆，进行了充分交流，获得了比较全面的认识。几年的参考、交流、考察，贵州生态博物馆给我印象深刻之处主要在于信息资料中心的建设。

贵州生态博物馆的建设规划，以梭戛为例，包括四个内容：1. 建立（长角）苗族信息中心；2. 对陇戛苗寨的原状保护；3. 建立组织机构；4. 财政安排。后来的三个生态博物馆均按照此框架进行建设。在四个内容中，重点是信息中心。“信息中心”由四个方面组成：一是信息库，其功能是记录和储存本社区特定的文化信息，包括口碑历史、文字资料、具有特殊意义的实物、文化普查资料等。二是参观中心，设立可供观众参观的小型展览。三是博物馆工作人员和志愿者的工作场所。四是社区服务场所，提供餐饮、会议室等服务。

笔者在2005年和2010年对梭戛、镇山、隆里、堂安四个生态博物馆的考察，重点都集中于信息资料中心。信息资料中心实际上是博物馆或资料馆，他们的建筑都很有地域或民族特色，展示内容为生产生活、节日庆典、工艺美术、宗教信仰等物质和非物质文化内容。信息中心的管理纳入地方政府文化和财政部门，设有专门管理、解说人员，所需经费由政府财政支出。有的信息资料中心设有客房，可供来访考察研究者住宿。贵州的四个生态博物馆都把信息中心的建设作为标志，信息中心一旦建成，即宣告开馆。如果按照前述国际生态博物馆的定义和内涵来看，所谓"建成开馆"其实只是开端，严格说只是"万里长征走完了第一步"，臻于健全和完善还有漫长的路要走。即使对照梭戛生态博物馆的整体规划来看，开馆也并非大功告成，才仅仅是部分基础设施的完工，距离设定目标还相差很远。对于这一点，建设者们也许心里有数，知道任重道远，而社会各界包括学术界却显得十分盲目，都把开端作为结果，忙不迭按照国际标准进行审视、评论、批评，于是乎好事变成了饱受质疑的对象。我们应该懂得，对于这样一项缺乏基础条件、牵扯面极广、需要长期投入智力和财力的事业，妄图立竿见影、一蹴而就，那是不可能的。别的不说，仅说欲实现生态博物馆的核心原则——民众参与做主，那就是一大难题，在这个问题上，即使再过 30 年，也难说能够达到目标。

尽管如此，贵州生态博物馆所产生的影响却不容忽视。1998 年 10 月 31 日，在梭戛苗族生态博物馆建成开馆仪式上，苏东海先生曾作了四点评估：1. 成功地在中国建立了第一座生态博物馆；2. 成功地创造了中国生态博物馆的模式；3. 梭戛生态博物馆在中国引起了广泛的注目；4. 唤起了梭戛人民对保护自身文化的巨大热情，激发了民族自豪感。国家文物局马自树副局长说：梭戛生态博物馆的建成开馆在中国博物馆领域具有里程碑意义，是中国博物馆专家学者对新博物馆学的勇敢探索和大胆尝试的结果（胡朝相，2011）。现在看来，上述总结显然是事业草创成功之时情不自禁的激动和喜悦的表达，对此我深为理解。在我们建设云南民族文化生态村之初，也有过自我感觉特别良好，并为取得一些初步成就而激动欢呼的经历。而当喧闹过后，不用别人提醒批评，发热的头脑冷静下来，一切又会回归于理智和现实。

受贵州的影响，内蒙、云南、广西、江苏、浙江等省区相继进行了生态博物馆建设的尝试。其中以广西的建设数量较多。2003 年，广西壮族自治区政府选择南丹里胡怀里（瑶族）、三江（侗族）和靖西旧州（壮族）作为试点建设民族生态博物馆。在取得一定经验的基础上，2005 年由广西民族博物馆编制《广西民族生态博物馆建设“十一五”规划及广西民族生态博物馆建设“1+10 工程”项目建议书》，并获得自治区民族民间文化保护工程领导小组的批准。广西的“1+10 工程”，即一个“龙头”博物馆——广西民族博物馆和 10 个民族生态博物馆的组合。10 个生态博物馆为前述 3 个馆加后来追加的 7 个馆：贺州市莲塘镇客家围屋生态博物馆、融水苗族生态博物馆、灵川县灵田乡长岗岭村汉族生态博物馆、那坡达文黑衣壮生态博物馆、东兴京族三岛生态博物馆、龙胜龙脊壮族生态博物馆和金秀县瑶族生态博物馆。笔者曾应邀参加过广西生态博物馆实施建设方案的研讨，并实地考察过龙脊、靖西旧州和那坡达文三个民族生态博物馆。广西的特点，主要创建了“1+10”博物馆体系，至于难点，同样是生态博物馆核心理念的落实。

云南也曾经学习贵州，进行了两个生态博物馆的设计和建设。一个是西双版纳傣族自治州勐海县西定乡章朗布朗族生态博物馆。此馆由西双版纳州委宣传部原部长黄映玲倡导并由州委宣传部拨款建设。章朗是一个典型的布朗族村寨，生态环境极好，传统文化深厚，民居全为干栏式建筑，错落有致，森林掩映，十分优美。村中有该区最为古老的佛寺，寺中的壁画被誉为南传上座部佛教瑰宝。2005 年，为建设生态博物馆，黄映玲部长亲赴北京请教苏东海先生，并拜访云南大学，希望我们给予协助。同年 10 月，在我们的帮助下，章朗生态博物馆开馆，苏东海先生应邀参加了开馆仪式。该馆建成后一段时间，主要依靠布朗族文化精英和村民自行管理，运作艰难。经努力争取，数年前已经被纳入政府文化部门管理体系，有了人员和维护经费的保障。由于理论、资料等准备不足，村民认知度不高等原因，该馆虽然开馆挂牌，但是许多工作没有跟上。2018 年 1 月下旬笔者再次造访章朗，恰逢寨里举行升佛仪式，宗教气氛浓郁，村民虔诚礼拜，民风古朴依然，只是被誉为“最美村落景观”的寨容已有破坏，传统民居未能坚持统一规划建设，正在无序改变，生态博物馆的宝贵优势逐渐削弱，令人遗憾！

云南的另一个生态博物馆建设没有章朗幸运，还未问世便“胎死腹中”。2004 年，云南普洱市（原思茅地区）孟连县主管文化旅游的副县长到昆明见我，告知该县富岩乡大曼糯寨尚保留着近百幢传统茅草房屋，景观稀罕壮观。村寨周围竹林茂密、古榕参天，苍翠欲滴；佤族习俗古朴，民风浓郁，能歌善舞，堪称“世外桃源”。县里的意见，希望把它建设成为生态博物馆，传承文化，保护生态，发展旅游，以改善佤族的生存状况。为此我两次前往考察，和县里达成合作协议后，组成了由建筑人类学博士施红和建筑学教授王冬负责、包括十余名研究生和本科生的调查、规划、设计课题组。经过两个多月的实地勘察研究，课题组完成了大曼糯生态博物馆的规划设计方案。方案通过专家组的评审和县里五套班子的审查，获得一致认可，县里决定立即实施建设。然而就在即将开工的前夕，项目突然中止，原因是省扶贫办公室指示孟连县在一个月内必须彻底拆除所有茅草房，否则会影响全县扶贫资金的下发。由于这样一个决定，使得大曼糯生态博物馆的美好计划成为泡影，使课题组几个月的辛勤劳动付之东流！目前在云南全省，完整保存着大规模草房景观的聚落只有一个——沧源县翁丁村。此村现为国家级非物质文化遗产“民族传统村寨”，还有“特色旅游村寨”等多个桂冠，旅游宣传则称其为“最后的原始部落”，每年慕名远道而来的国内外旅游者络绎不绝，业已成为云南靓丽的旅游名片之一。看今日翁丁村草房部落风光无限，令人对当年匆忙消灭大曼糯寨茅草房、终止该村生态博物馆建设的举动感到惋惜！

四、博物馆人类学的视野

介绍了国内外生态博物馆之后，人们不禁要提出一个问题：生态博物馆应属于哪个学科范畴？回答也许是肯定的，那就是博物馆学和博物馆事业范畴。为什么？因为国内外的生态博物馆建设概无例外均由博物馆学家倡导和创建，他们都把生态博物馆视为博物馆事业发展的一项创新模式或称“新博物馆学运动”。不过，从生态博物馆的理论和实践来看，它显然已经打破了传统博物馆的边界，大大超出了传统博物馆学和博物馆事业的范畴，被赋予了其他学科尤其是人类学的丰富内涵。关于这一点，不妨从本文第二节学者们对“什么是生态博物馆”所下的定义来看，就不难明白。上述定义十分清楚地彰显了生态博物馆不同于传统博物馆的三个鲜明特点：进行自然遗产和文化遗产的就地保存、培育、展示；由住民参与，由行政当局和住民共同构想、创造、利用；有助于地域社会的发展。这三个特点其实就是人类学理念的表达。

如果定义还不够具体完整，请再看挪威专家与国内学者共同制定的贵州生态博物馆《六枝原则》。《六枝原则》共九条：1. 村民是文化的真正拥有者，他们有权利按照自己的意愿去解释和认同他们的文化。2. 文化的含义与价值只有与人发生联系并依据自己的知识得以界定和解释，文化的内涵才得以加强。3. 生态博物馆的核心是公众参与，文化是一种共同的和民主的构造，必须以民主方式加以管理。4. 当旅游业与文化保护发生冲突时，后者必须给予优先权。原件的文物是不应该出售的，但以传统工艺为基础的高质量的纪念品生产应该得到鼓励。5. 长期的和历史的规划是至关重要的，必须避免短期经济利益损害文化和长期利益。6. 文化遗产保护必须融入整体环境，传统技术和物质文化资料是核心。7. 观众有道德上的义务和尊重的态度遵守一定的行为准则。8. 生态博物馆没有固定的模式，因文化和社会条件的不同而千差万别。9. 社会发展在生态博物馆的建设中是一个先导条件，人们生活的改善必须得到更多的重视，但不能以损害文化价值为代价（胡朝相，2001）。

看过《六枝原则》，感觉就是一个纯粹的“博物馆人类学宣言”，它通篇体现着人类学的基本通则和核心价值，远远超出了生态博物馆的范畴，对于地域和民族文化的认知、研究、保护、传承、发展，对于乡村振兴等，均有指导意义。可以预言，从长远看，包括贵州在内的生态博物馆的贡献，并不在于作为基础设施的资料中心等的建设，那些设施若干年后必然更替淘汰，不会留下多少痕迹，而生态博物馆的理念如《六枝原则》等，却会越来越受到人们的认同和重视。仅以乡村建设为例，和十余年前相比，现在可谓形势大好，各种建设、振兴、活化、孵化的战略、理论、思路、途径、方法不胜枚举，异彩纷呈。然而如果从文化理论层面看，却没有一个可以和《六枝原则》相提并论，没有一个可以和《六枝原则》的理性、人性、高度和深度相媲美。何以如此，关键所在，就是人类学基本通则、伦理、观点的不足和缺失。

然而问题在于，上述观点要让学界和社会接受尚需时日。迄今为止，对于《六枝原则》等生态博物馆的理念赞赏者很少，而责难否定者却较多，如认为“太过超前”、“太过理想”、“太过乌托邦”、“不适合中国国情”等。20 世纪 80 年代到 90 年代，全世界曾经出现了 300 多座生态博物馆。从欧洲向北美、南美、澳洲传播和发展，但是到了 20 世纪 90 年代之后，消失了将近一半。说明生态博物馆的理论和实践、博物馆学和人类学的结合并非易事。对此，苏东海等进行了如下反思和回应：首先，引进生态博物馆带有一定的盲目性。他曾坦言：“我那时候头脑发热。在‘七五’规划的时候，我就提出：贵州只有一个省馆，要想在全国后来居上，怎么办？我提出要建设生态博物馆，但我也不知道怎么搞。我在'国际博协'的年会上，接触了一些新博物馆运动的学者，了解了他们的想法和做法。这样我就将自己的想法落实到实践上去"（苏东海，2017）。其次，建设生态博物馆的困难在于先进超前文化与原始古老农业文化的不相适应。苏东海在反思中多次说到，生态博物馆乃是欧洲后工业社会的理念和想法，是欧洲文化中比较先进、超前的文化。而梭戛等民族村寨还处于农业自然经济的阶段，两者时空距离差得很远，要把先进的思维嫁接、移植到古老的农业文化之中，必然产生碰撞，甚至难以存活。三是提出"文化代理"概念。苏东海认为，建设生态博物馆，要使外来先进文化嫁接到本土文化上，需要一个政府和专家操作的文化代理阶段。因为当地村民并没有这个要求，是政府和专家积极热情地强加给他们的，所以如果没有文化代理阶段，外来文化引进和嫁接就不可能实现。什么时候代理阶段结束了，就差不多达到目的了。

再者，生态博物馆成功的难度，在于从文化代理回归到文化自主，村民需要经过三个文化递升层面。苏东海解释说："建立生态博物馆，政府是积极的，博物馆专家的热情也很高，村民由于利益的驱动，也是积极参加的。在这三方面，专家和地方干部是主导力量，村民是被领导，因为他们并不知道什么是生态博物馆，也不知道要干什么，我不得不说，事实上外来的力量成了村寨文化的代理人，村民则从事实上的主人变成了名义上的主人，没有外来力量的进入，就不可能有生态博物馆。这是事实，也许别的国家不是这样，但中国是这样。在中国建立一个生态博物馆并不难，而巩固它比建立它难多了，因为建立它是政府和专家的行为，而巩固它只有文化主导权回归到村民手中，村民从名义上的主人回归到事实上的主人时，生态博物馆才能得以巩固。生态博物馆的核心理念在于在文化的原生地保护文化，并且由文化的主人自己保护，只有文化的主人真正成为事实上的主人的时候，生态博物馆才能巩固下去。也许外国在那些文化程度高的地方建立生态博物馆不需要别人的帮助，而中国确实存在着文化代理阶段。从文化代理回归到文化自主，村民需要经过三个文化递升层面，这就是利益驱动层面、情感驱动层面和知识驱动层面。村民保护自己文化的动力来自利益的驱动，来自对自己文化的天然感情，来自对自己的文化价值的科学认识。”（苏东海，2005）

上述反思不仅是生态博物馆经验教训的总结，也应该是所有人类学民族学博物馆以及博物馆人类学值得深入思考研究的问题。我国人类学民族学关于“物”的收集、研究、展示，以中央研究院开创的民族学事业为标志，滥觞于20世纪20、30年代。1949年中华人民共和国成立后，随着各地少数民族自治州和民族院校的建立，一批中小型的民族学人类学博物馆相继问世。20世纪90年代以后，以云南民族博物馆建设为先导，一批现代化的大型民族博物馆陆续诞生于各民族省区。最近几年在国家民委的领导下，每个少数民族建设一个博物馆，目前此项工程已近完成。经过将近一个世纪的发展，我国的民族博物馆无论数量还是规模，均已蔚为大观，走到了世界的前列。

然而，令人遗憾的是，事业虽然蓬勃发展了，理论研究却明显没有跟上，博物馆人类学的研究几乎是空白。所有人类学民族学博物馆遵循和应用的理论和方法，基本上是传统常规博物馆的理论和方法，鉴别分类的概念和标准均以考古学文物学为依据，民族学资料缺少自身独立的阐释，多半被作为考古和历史文物起源演变的佐证，这无疑削弱了人类学民族学博物馆的特色，制约了其特殊文化功能的发挥。据上可知，生态博物馆在我国的发展虽然远远不能与常规民族博物馆相比，然而却有其突出的优点。优点一，理论方法先行。国外有明确的定义概念和理论方法体系，国内则有《六枝原则》等理论的开发。优点二，具有超前理念。如“文化原生地保护”、“村民是文化的真正拥有者”、“公众参与”、“文化遗产保护融入整体环境”等。优点三，本土化追求。如提出“文化嫁接”、“文化代理”、“政府专家主导”、“文化回归”、“文化递升三层面”等根据本土经验提炼上升的系列思想和学术概念。上述理论、思想和概念，有的已经具有普世价值，有的虽然尚不成熟完善，甚至有缺陷，存在质疑、商榷、批评的空间，然而学术的真正价值不在于人云亦云循规蹈矩，而在于标新立异开拓创新，哪怕幼稚另类。如果说生态博物馆研究于博物馆人类学有所裨益，笔者认为就在于上述三点。博物馆人类学的基础研究，完全可以从这三个优点的具体内容着手深入探讨，进而建构厚实和宽广的理论体系。

参考文献

大原一兴：

《生态博物馆之旅》（日文版），（日）鹿岛出版社，1999 年 12 月。

广西壮族自治区文化厅：

《广西民族生态博物馆“1+10 工程”建设项目》资料集，2005 年 8 月。

胡朝相：

《论生态博物馆社区的文化遗产保护》，载《中国博物馆》，2001（4），19-22。

《贵州生态博物馆纪实》，中央民族大学出版社，2011 年 10 月。

黄春雨：

《中国生态博物馆生存与发展思考》，载《中国博物馆》，2001（3），2-19。

李志玲译：

《国际博协关于博物馆和文化旅游原则声明的提案》，载《中国博物馆通讯》，2001 年 8 月。

孟凡行等：

《生态博物馆建设与民族文化发展——以梭戛生态博物馆为中心的讨论》，载《原生态民族文化学刊》，2017，9（4）128-140。

苏东海：

《生态博物馆在中国的本土化》，载《中国文物报》，1999 年 3 月 28 日。

尹绍亭：
《民族文化生态村：理论与方法》，云南大学出版社，2008 年 11 月。
尹绍亭主编：
《民族文化生态村——云南试点报告》，云南民族出版社，2002 年 10 月。
中国博物馆学会编：
《2005 年贵州生态博物馆国际论坛论文集》，紫禁城出版社，2006 年 2 月。

青年新作

从“二月二”到“还炮节”：一个华南村落的仪式再造与社会再生产

谢杲馥*

节庆与仪式能够集中地表现文化意义，因此一直是人类学研究的偏好所在。不过，学界关于“二月二”的研究并不全面，大多是从民俗、节俗的角度入手，或者是考问节日由来和文化传统。另外，近十年来，将“二月二”作为地方非物质文化遗产进行研究的相对较多。笔者最初也正是因为这样的原因才考察了广东省罗仁村的“二月二”。因为在官方的说法里，罗仁村的“二月二”隶属于河台镇的“开耕节”，是2013年获批的第五批广东省级非物质文化遗产“高要春社”的重要组成部分。不过，随着调查的深入，笔者发现罗仁村现在过得如此隆重的“二月二”在村民的认知里叫作“还炮节”，而且这个叫法是在近十多年的活动中逐渐固定下来的，现在这个节日是全村人一年中最重要的狂欢时刻。它更多地呈现出仪式恢复社会平衡、巩固群体团结、完成社区界定和推动社会再生产的意义（彭文斌、郭建勋2010）。

* 谢杲馥，肇庆学院旅游与历史文化学院。

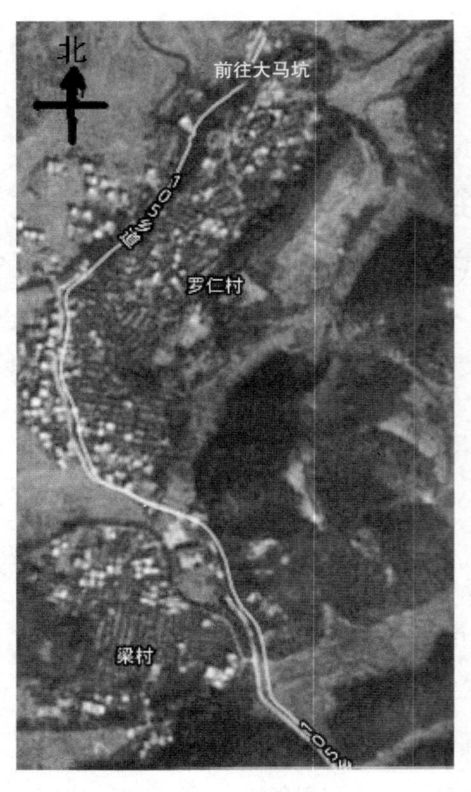

图1：罗仁村鸟瞰图

一、田野点介绍

罗仁村是广东省肇庆市高要区河台镇辖内的一个行政村，位于河台镇的西南角，离镇中心约有10公里。下辖梁村、罗仁村、大乌坑三个自然村，其中罗仁村是人口最多、占地最广的。而且形似贝壳，由一岭一路怀抱。岭是被称为“大埠岭”的小山丘，位于村子东边，其上有一块大约一个足球场大的平地，是每年“还炮节”的主会场，因此被称为“炮台”。而路则是大村前那条南北走向的105乡道，该乡道将三个自然村相连。不过大乌坑离得较远。而梁村与罗仁村仅一路之隔，位于罗仁村的西南偏西方向，当地人把罗仁村和梁村分别称为上村与下村。[1]（见图1）

罗仁村只有焦、梁两个姓氏。总人口数约1950人。其中，最大的上村（罗仁村）约1200人，大部分是焦姓，仅有5户是梁姓。下村（梁村）都是梁姓，约450人。大乌坑村与这两个村相隔比较远，且四面环山，交通相对不便，约300人，都是焦姓。

据焦氏族谱[2]记载，焦氏祖先在明末迁至广东高要新桥。之后四子焦徐杨携妻王氏又从新桥迁至罗仁村。由此，罗仁村自始祖焦徐杨传至今日已有十七代。[3]另外，族谱中记载，大乌坑乃焦氏分支。因清末时田地紧张，族人焦喜朝举家迁至罗仁村的东北向约两公里处的小山坳大乌坑定居。另外，梁村和上村（罗仁村）里的梁姓都来自于同一个梁氏家族。据梁氏族谱[4]记载，梁氏祖先居于罗定，五子演志迁居罗仁下村，传至今也有十五代人。从两姓的族谱来看，梁姓和焦姓的联姻很常见，宗亲关系密切。另外，与其他村落不同，现在罗仁村的民众并不以房支来区分人群，而是以中队划分群体。整个罗仁行政村分为五个中队，大乌坑为一中队，二、三、四中队在上村，第五个中队是梁村。

最后来看村中两类重要的公共空间——书室与大榕树。这也是罗仁村与其他村落不同的地方。首先，与高要地区其他村落喜欢建“酒堂”、翻新祠堂不同，罗仁村虽然近年来活动越办越多，但并没有新建“酒堂”，也没有大肆翻新祠堂。而且村中三处祠堂，村人也都习惯于以“书室”称呼。其中最大的书室——学周书室，位于上村中南部，是焦姓的总祠堂，又名“以忠焦公祠”，始建年月不详，但族谱中记载民国36年进行过一次翻修。2008年村人又进行了一次更换门楣和祖先牌位的小规模翻修。第二间书室在上村西北部，叫愷元书室。村人的说法是这间叫二中队祠堂，三、四中队并不过来祭拜。因此在笔者看来，这应该是焦姓的分支祠堂，而且多方查证后笔者发现此祠堂建立时间与大乌坑建村时间点相合，想来彼时的焦氏应该经历过一次重大的分家，不过此问题笔者将另行讨论，在此并不赘述。第三间书室名为崇德书室，位于梁村东部，其实就是梁氏祠堂。另外，作为焦姓分支的大乌坑没有祠堂，仅有一间主屋。

除了祠堂特殊外，村中的三棵大榕树也有其独特之处。在南方的村落里，村口大榕树一向是村落重要的公共空间，村人喜欢在榕树下纳凉聊天，甚至集会。不过罗仁村的这三棵大榕树平时并没有人聚会聊天，因为每棵榕树下都有一个小小的土主庙。这三座庙显然都是新建的，不过笔者询问过多位村中老人，他们均表示，榕树下的土主庙是罗仁村的传统，即使在建国后很长一段时间里土主庙被毁了，但是村人仍然会去那里拜土主。有少数的村民还会让孩子认土主为“干亲”。在二月二当天，龙母环村大游行中拜土主也是第一个重要的环节。这三个土主庙分别在上村的“村顶”（村子正北方）、炮台下方（大埠岭东南方向）和梁村的偏南方（见图 2）。另外，大乌坑有自己的土主，在村子的东、西边各有一个。不过，大乌坑并不参加罗仁村的节日仪式。

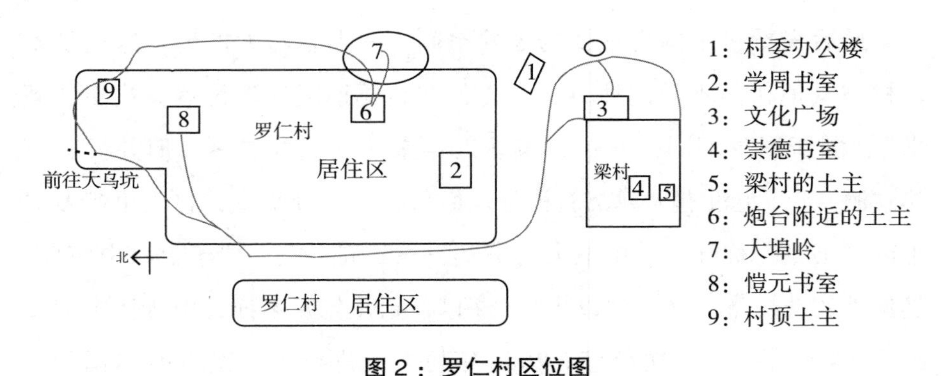

图 2：罗仁村区位图

二、节日的复兴

笔者原本是带着考察非物质文化遗产的目的进入罗仁村的，没想到刚刚转完村落就发现了此村与高要其他村落的一些不同，通过进一步参与村落社会生活，发现上述三种特别之处恰恰都与罗仁村的还炮节相关。而且罗仁村的“还炮节”还有一个节庆复兴与仪式再造的过程。此一过程恰与中国历史社会的发展脉络相关，也呈现出地方社会的内在组织性，节日组织的大小、强弱、盛衰代表着乡土社会本身的活力，也代表着乡土社会与外部大世界的关系（高丙中 2016：52）。接下来，笔者将梳理一下罗仁村“还炮节”的历史，并试图分析节日复兴过程中地方社会组织是如何变化和展示的。

1. 民国时期

还炮节在罗仁村村民的记忆中由来已久，但谁也说不清楚这一节日到底源于何时。不过村中九十岁以上的老人[5]回忆说，民国时罗仁的"二月二"也是非常热闹的。那时更多地称为"龙母节"，也有人叫它"龙母花炮节"、"龙母还愿节"。当时节日的主要内容就是龙母环村大游行和燃放"龙母丁财炮"的仪式。据说那时罗仁村的"二月二"是各家各户先自行拜祭祖先，再聚集祠堂进行巡游。巡游过程是先拜土主，再去龙母庙拜祭龙母。[6]最后，村民会回到大埠岭燃放龙母丁财炮。这一流程跟今天的还炮节颇为类似。不过，有两点值得注意的不同：一是，梁村的一位老人反复跟我强调的，那时候梁村是单独进行拜祭土主、龙母和放炮仪式的。而且因为都需要在龙母庙祭拜龙母，所以每年两个村都会提前一天用抓阄的方式决定祭拜龙母的顺序和时间。大乌坑也是自己祭拜，但是他们不会回来祭拜龙母，仅祭拜祖先和土主，而且没有放炮仪式。二是那时候没有炮会，活动由地主组织，不过地主会组织"太公会"来负责龙母节的活动。因此游行的队伍不仅会抬龙母牌位，还会抬地主和族长游行，民众还需要向他们行礼磕头。另外，地主们为了添丁发财，会放很多炮，持续很长时间，也是热闹非凡。

2. 20 世纪 50—80 年代

1949 年后，没有了地主的组织，罗仁村的"二月二"虽然不那么热闹，但还算是井然有序。那时候人们更多地倾向于先在家隆重地祭拜祖先，再去土主庙和龙母庙上香。然后族长和族老会组织一些人去大埠岭上放炮庆祝。

到 20 世纪 50 年代末，人民公社化运动时期，罗仁村的"二月二"节庆受到管理，不再有族长、族老带领下的放炮活动了。然后到了

“文化大革命”时期，“破四旧”开始，村里的祠堂、神像、神台、祖先牌位都被不同程度地摧毁了，龙母神庙和土主庙更是被夷为平地。在家里祭拜祖先也不被允许。也正是从那时起，这一天不再叫“龙母节”，而是直接叫“二月二”或者“双二节”。这样的状态一直持续到上世纪70年代末期，经历过“文革”的动乱，大家都将龙母信仰当作封建迷信，不敢再提。这时候四中队中有一个三代单传的大龄单身汉JZD，[7]有一次在水渠网鱼，居然找到了被红卫兵丢弃的龙母牌位。[8]他见牌位完好无损，不顾旁人的劝阻，带回家中供奉。结果当年就娶了妻子生了孩子，而且在以后的三年里，又连生了两个男孩，几个孩子也都很有出息。这个事就此在村里传为佳话，大家都说是龙母显了灵，也因此人们开始想重启龙母节的活动。但1982年当时的乡长向上级建议再次集体举办龙母节，却被说是想搞封建迷信，受了批评，也未能如愿，就此打击了民众的信心。

3. 90年代—21世纪

一直到1989年，政策逐渐开放，社会也越来越自由，罗仁村周边地区的庙宇、神坛和传统节日都逐渐恢复。那一年的“二月二”，在全村人的强烈要求下，JZD先生把龙母牌位重新拿出来进行集体供奉，拜土主、拜龙母的节日仪式也得以重新举办，村里还以各中队为单位，抽出人手组织了炮会，专门负责燃放丁财炮，而且定下了以后每年以中队轮值炮会的规则。不过，节日恢复初期，活动并不隆重，巡游活动最开始也没有恢复。[9]但是，值得注意的是，据老村长的回忆，从这次“二月二”活动重启开始，上村就和下村一起祭拜土主和游行了，因为是以中队的形式组织的炮会。而且为了显示友好，焦姓族人还特意将龙母巡游的路线定为先去祭拜下村土主，再来祭拜上村土主。另外，据说是按照传统惯例，巡游依然没有前往大乌坑。此后大约十年的时间内，罗仁村为了是否重建龙母神庙争论不休，虽然每年都有炮会来组织二月二的活动，但活动都不够大，也不热闹，村民们也一直管这一天叫双二节，而不是龙母节。

4. 2002 年以后

这样的状况终于在2002年有了改变，这一时期罗仁村进入了现代化冲击传统村落的阶段。当时的广东沿海地区正是中国万千青年打工淘金的圣地，罗仁村的青年们也纷纷走出村子。这就使得罗仁村的双二节陷入了可以办、但没人参加的尴尬境地。也正是这一年，三中队的JXR先生担任了罗仁村的村支书。据他自述，当时感慨于村庄过完年之后太冷清，也怀念自己小时候热闹的龙母节，他决定要好好举办这一年的二月二节日仪式。当年当值炮会的刚好是三中队，JXR先生就联合队中几位老人，改良了二月二的活动。第一，他在节日前事先联系了在外面发展得比较好的几位自家亲戚，并说服他们出钱给村里办二月二。然后他又将三中队外出不远的年轻人都喊了回来，协助他举办二月二的活动，所谓“有钱出钱，有力出力”。第二，他首次在大埠岭的炮台处搭建了表演舞台，并请来民间文艺团体在祭拜龙母后进行文艺表演，给节日造势。[10]第三，也是最重要的一个变革，他和队里的几个元老商量决定将他外家德庆地区的抢花炮活动融入到放龙母丁财炮的过程中。他将龙母牌位底座的铜环作为炮箍绑在土炮上，在文艺表演后，上台讲话并向大家展示铜炮和炮箍，然后宣布抢炮箍的事情。那一年因为新颖的文艺表演吸引了不少村中的年轻人，所以随之而来的抢炮箍活动就成了年轻人的狂欢。更有趣的是那一年抢到炮箍的四中队小伙JLY当年就生了个白白胖胖的大儿子，而四中队在那一年还诞生了罗仁村第一个名牌大学的大学生。村民们更是说，那一年四中队家家户户都蛮顺利，生儿子的家庭也相对多一点。于是顺理成章地，抢到炮箍的人及其所在的中队下一年就成为了炮会的组织者。而且据说是因为炮箍会不断被还回去，所以这个节日的名称逐渐统一叫做“还炮节”了。[11]

JXR先生的创新活动使得罗仁村的“二月二”再次成为人们讨论的焦点。直到今天讲起这年的抢炮活动，很多村人依然津津乐道，就此固定了罗仁村还炮节节庆活动的基本模式。即先巡游拜土主和龙母，再进行文艺表演，最后进入放炮抢炮箍活动。更值得一提的是，此后每一年轮值的中队形成竞争状态，不断给还炮节增添新内容。2003年增加了龙母巡游过程中的龙母塑像，并完善了游神队伍。2004年增加了各家燃放添丁炮的部分。[12] 2005年又为这个节日增加了一个起炮福活动，不过不是在二月二当天举行，而是抢到炮箍的中队在两周内择吉日举办聚会，正式遴选成员加入炮会，推举炮会总负责人，并开始为期一年的筹款、联络、组织等活动，以保证来年二月二炮会越办越隆重。2006年，又增加了还炮福，也是一个炮会的商议活动。是在还炮节两周前召集炮会成员进行聚餐，并就当年的还炮节活动进行细致安排。

除了这些大的变动，其余小的变化几乎每一年都有，几个中队形成了竞争模式，把活动越办越隆重，罗仁村的还炮节也由此出了名。越来越热闹的节日渐渐感染了周围的村庄，人们开始一传十、十传百地进行宣传。到如今，罗仁村的还炮节已成为高要地区最有名的民间节日之一了。特别是2012年开始，高要区政府投入大量资金支持河台镇“开耕节”[13]的举办，吸引了各界人员前来参观游玩，罗仁更是成为焦点，吸引了多家新闻媒体的访问报道。现在，每一年的还炮节都有很多外来游客或摄影团队驻足，甚至还有外乡人会为炮会捐款筹办还炮节。[14]

以上是罗仁村还炮节节日复兴的四个阶段。我们大致可以看出，在传统社会时期，这个节日由地主和地方精英主导，这些人有钱出钱，有力出力，承担了这样一件公共事务。建国后，我们为了进入现代化国家，在国家层面上逐步抛弃了“旧文化、旧风俗”，也因此摧毁了传统的地方社会和基层组织（徐天基 2013）。更因为十年“文革”的高压文化管理政策，使得后来即使社会变化了，民众仍不敢接续传统。一直到改革开放七年以后，活动才得以恢复，但在很长一段时间内这种活动与政府、国家间意识形态的紧张感并未消除。21 世纪初，国家改革措施深入，各种新的社会力量成长，加之新的经济社会形势带来农村社会的变化，罗仁村的还炮节才真正复兴。这一时期，村落社会受到现代化的冲击，村人逐渐外出务工，村落面临空心化、老龄化状态，外出的村人之间也不再熟悉，所以还炮节活动就是一个地方性的公共事件，由共同的信仰、共同的习俗把众多的人事组织起来，构成一个相互协作的仪式过程，满足各种人群与组织的相同和不同的需求与目标，也正是这一过程维持了地方社会的再生产（高丙中 2017）。至于它是如何积累和分配公共性，完成社会再生产的，我们通过节庆仪式过程再来进行分析。

三、仪式的再造

笔者从 2013 年开始，连续跟踪了五年的罗仁村还炮节。[15] 通过对节日仪式的参与观察，发现罗仁村的还炮节已发展成一个当地民众的狂欢节，平日里外出打工的年轻人都会在这一天回到家乡参加还炮节的抢炮活动。同时，他们也通过参加炮会组织完成自己村落身份的构建。接下来，我们具体看看还炮节的仪式流程。

1. 祭祖

还炮节当天的第一个仪式就是祭祖。不过，这个活动是村民们在家中各自完成的。村中老人说在传统社会时期，祭祖是非常隆重的。但今天看来，祭祖已非常简单，仅仅是家中妇女在堂屋或是祖先的照片前摆上准备好的鸡（鹅）、苹果、糖果、饼干、煎堆、茶蛋、茶和酒作为贡品，家人起床后上香拜拜即可。很多家中的男人一早起来就去村里给炮会组织帮忙，仅仅草草点上一炷香就算完成祭祖了。唯一特殊的是，当年的炮头（上一年抢到炮箍的人）在祭祖前要先去书室（祠堂）供奉龙母的地方祭拜龙母，然后再回来祭拜自己的祖先，祭拜完后又要将贡品端回到龙母神像前进行供奉，等到“龙母环村大游行”仪式时，炮头要派一个人（一般是自己的父亲）端着这些贡品进行巡游。另外，还炮节当天，外嫁的女儿都会回到娘家，尤其是这些年活动越办越热闹，很多人家要来不少亲戚朋友，而且大家为了看村里热闹的巡游，差不多在各家刚刚祭完祖就到了。所以，这几年村中就统一将“龙母大巡游”的时间定在上午十点半开始。这之前就是各家各户接待亲戚的时间。显然，在这一天的仪式中，集体的活动远比家庭的仪式来得重要。

2. 龙母环村大巡游

九点半左右，村里的居民（尤其是男人们）不管是不是炮会组织的成员，大家都开始去往供奉龙母的那个书室。然后，大约在十点，炮头、炮头的父亲和几个族老以及当年的炮会负责人就会进行一个请龙母仪式。这其实是游行前最重要的准备工作。先是当年的炮会负责人念念有词，然后炮头点香叩拜，最后几人一起拿柚子水擦拭龙母牌位、神像与其神龛。仪式并没有专职的神职人员操作，整个过程也很简单，不过却很庄重。仪式开始后，人声鼎沸的书室会立刻安静下来。仪式完成后，炮头会戴上红色大胸花，并着力保护龙母牌位，实际上就是防止龙母牌位下的炮箍被人摸或者被人抢。这一两年更是为了增加炮箍的神秘感，将龙母牌位放在一个托盘上，并盖上了红布。与这个过程同时，炮会的其他成员会有序地准备好游行的各种需要，并基本上排好行走的顺序。最后抬来烧猪，点燃三支大圣香再次向龙母上香祭拜。然后准时在十点半开始敲锣打鼓，舞起狮子，请龙母出游。每年看这个过程都让笔者感叹，炮会组织的行之有效和炮会组织在这个过程中的重要性和权威性。

巡游开始后，队伍的顺序是：舞狮队在前、然后跟着鼓乐队、再来是炮头端着牌位、童男童女四人抬起龙母神龛、炮头的父亲端着刚刚祭拜龙母的祭品、炮会成员抬着烧猪，以上是主要部分。后面就跟着炮会工作人员、保安队和村里看热闹的人，这几年队伍越来越长，人越来越多。大部分的村里人都会跟着队伍游行，说是一起向龙母祈福。

游行的路线是先从当年供奉龙母牌位的书室出发，一路行至梁村大榕树祭拜土主。祭拜土主的仪式有点特别，首先是摆上贡品，接着点香祭拜，然后还要由舞狮子进行拜神（绕土主庙转三圈，并且叩首）。在大队伍离开时，梁村会有一个负责人（一般也是当年炮会的成员，但都是梁村人）在土主庙处燃放提前准备好的长达十几米的长炮。这个仪式结束后，队伍就来到村委办公楼左前方的一棵大樟树（曾经龙母庙的位置）进行简单拜祭，也以放炮收尾。然后主要的游行队伍会到文化广场休息整顿 20 分钟。之后游行队伍再次出发，缓慢地在村道上绕行，这时候的队伍会加上更多人。下午要进行文艺表演的队伍也会跟在主要队伍之后，然而这几年为了增加节日气氛，每一年的炮会组织会请不同的队伍来撑场面。比如 2013 年请了龙旗队，2016 年请了龙旗队和执伞队一起，2017 年又增加请了德庆地区的五龙子（扮演龙母的五个儿子的表演队），2018 年更是从广宁请来的十八堂狮子，队伍可谓是浩浩荡荡。当队伍经过居民家门口，或者离得比较近的时候，村民就都会在自家门口燃放鞭炮，以示向龙母祈福，队伍也会停下来，接受炮礼，并派舞狮去村民家门口转一转。这样队伍一路爬坡，走到村顶的土主庙进行拜祭。然后再继续游行到炮台处祭拜最后一个土主庙。到此龙母环村大游行就结束了。

在人类学上，我们都知道游神的意义是为了划定边界，彰显权威性。罗仁村的游神活动也不例外，很显然在村人的心目中，异姓的上村和下村是一个整体，但同姓同宗大乌坑居然被排除了出去。至于原因，笔者尚未得到确切的答案。另外，罗仁村的龙母大巡游是每年炮会组织者最爱粉饰的部分，尤其喜欢通过请表演方队与外界社会进行联系，试图展示自己的村落文化，当然我们也可以将这看作地方精英处理与外界社会关系的一种方法，只是略显简单了一点。

3. 文艺表演

接下来，经过短暂的休息，下午一点半左右，全体人员会再次集合到炮台，先是由请来的表演队表演，村民们会陆陆续续过来观看。表演一般都紧跟时代的步伐，也都是现代性的居多，大约会持续一个小时。然后不管每年请来的表演团队是什么，最后都一定会有一个狮子上高桩的节目。舞狮队就是早晨游行时的狮队，一般是村里的年轻人来扮演的，这几年为了壮大队伍，也会从外地请一些，但是村里的队伍一直在。这个节目是有一定难度的，首先需要几头舞狮分别登上一个插了高杆的圆木桩，舞狮要用嘴巴咬掉高杆上的捆扎好的大葱和芹菜。全都咬掉后，炮会有人放一串很长的鞭炮，表演部分就结束了。笔者访谈得知，舞狮表演是村长 JTC 当值炮会负责人的时候定下的，我询问老先生，他说就是希望村里有人继续学习舞狮，不要把这门手艺丢了。这让我很感叹，地方精英对于文化的保护是非常用心的。不论是 JTC 先生还是 JXR 先生，这么多年都默默地为还炮节服务，即使不是他们自己的中队当值，他们也会主动在炮会里服务。还有一个JZY先生，他是村里的老会计，每年都会喊自己在外当老板的二儿子捐很多钱。笔者问他为什么，他说也不是单纯想着得龙母保佑，就是觉得这个活动是村里集体的事，有能力就应该多做贡献。另外，他的儿子太忙，不常回来帮忙，[16]老人觉得应该多多捐钱才能弥补他不为村里服务的遗憾。在访谈这几位年老的地方精英时，他们普遍表现出对村落集体的高度认同和维护，而且非常爱惜村里这样的状态。笔者以为目前罗仁村还炮节的发展和罗仁村社会的良性发展正是与地方精英的良好品格相关，而且这些精英是一代人。

4. 拜龙母

在舞狮结束后放鞭炮的过程中，文艺表演队撤下舞台，炮会组织会迅速布置龙母香案，然后由炮头请上龙母牌位和神龛。接下来就是拜龙母的时间了。不过不是全村人都拜。这个祭拜过程是由炮会负责组织的，他会先请上本年度为炮会捐款最多的十个人，给龙母上香并祭拜。然后再请上上一年添丁的人家挨个上香。再然后是上一年抢到龙母丁财炮炮箍的炮头上香，最后开放时间允许村人自由上香。这里需要说明的是，请捐款前十名上香的活动是最近才发展出来的，也是由老村长JTC先生提议设立的。访谈时他告诉笔者这是自己从德庆龙母庙会学来的，主要是为了表彰给村里捐钱的人。这几年活动做大了，还有村外人愿意给炮会捐钱，炮会也会邀请他们来参加还炮节，并允许他们给龙母上香。笔者询问理由，村里人都说龙母娘娘会保佑好人，而且龙母娘娘是大众的神仙，不能阻止别人进奉龙母娘娘。这正表现了当地人对龙母信仰的认知，而且这个认知是地方性的，村里人、村外人都有这样的表现。

5. 放添丁炮并抢炮箍

所有人祭拜完龙母后，上年添丁的人家在炮会的组织下，会聚集到舞台前的空地上，挨个燃放添丁炮。添丁炮是由炮会发放的，每一家都一样，是一个有炮座的土炮。从2004年开始，添丁炮也绑上一个铁炮箍，可以抢炮了。而且抢到炮箍的人，燃放添丁炮的主人家会送上红包与糖饼。有些还会请回去一起吃饭，寓意着把自己上年的好运传递给对方，互相祝福丁财两旺、万事如意。很显然，这个过程带着明显的礼物流动的意义，是典型的社会再生产的交换活动，有互惠和再分配的过程。需要注意的有两点，一是从上村和下村一起举办二月二活动开始，丁财炮就是一起放的。2004年恢复各家各户放添丁炮，梁村的添丁户也是跟着一起放添丁炮的，这个传统一直延续着。这进一步说明了上村和下村是一个完整的村落结构。第二点是这个活动从2004年才开始，但很受村里人的欢迎。原因是大家外出的多了，谁家这一年添丁进口也不一定知道，有了这个活动，大家能够了解一下村里的新动态。而且因为派红包和请吃饭，多了关系往来，感觉更熟悉一点，不会因为外出而陌生。这就说明这个仪式的改造是地方社会为了应对现代化社会模式冲击而进行的。

6. 放龙母丁财炮及抢炮箍活动

添丁炮的抢炮箍活动已经把村子的气氛调动了起来。添丁炮放完以后，紧接着就是最大的龙母丁财炮的燃放和抢炮箍活动了。首先由炮头叩拜龙母，然后把龙母牌位底座的铜环拿下来，并由炮会人员绑在即将燃放的龙母丁财炮上。龙母丁财炮也是土炮，但是比添丁炮大很多，这几年都是二十响的。是专门找人做的，而且龙母丁财炮是铜炮座，且比添丁炮的炮座大很多。这时候已经差不多下午四点了，龙母丁财炮很快就会放完，炮箍就会被炸飞。于是狂欢开始了。众人便蜂拥而上，争抢丁财炮。因为是以抢到炮的中队来组织炮会，因此抢炮活动也演变成了以中队为单位的竞技活动，整个中队的年轻人会齐心协力去完成这个过程。而且这几年各中队还会提前集合成员进行任务的安排，跟打篮球比赛一样，居然有防守掩护的和冲锋陷阵的专门人员，而且拿到炮箍的人还会被中队成员牢牢地保护起来，一直到他走上舞台把炮箍放回龙母牌位下，这个活动才算正式结束。抢到炮箍的中队会欢天喜地、敲锣打鼓、舞着狮子，迎回龙母牌位和龙母神龛，放在自己中队的书室里。中队里的妇女和孩子们这时候可以上前叩拜龙母，而男人们则一边在书室门口放炮庆祝，一边热议刚刚那一场激动人心的抢炮箍活动。村里的很多妇女还告诉我，这一年抢到炮箍的中队男人们都特别神气，总是不断谈起自己抢炮箍时是何等英勇。

7. 起炮福与还炮福

上述为罗仁村还炮节当天的仪式。与之相关的还有两个仪式——起炮福和还炮福。这两个活动是从 2005 年开始，但不是全村性质的，只需要炮会成员参加，完全可以看成是还炮节活动的两次筹备会。起炮福是在还炮节后的两周之内择吉日进行的，主要目的是确立新一年炮会的成员，并推选当年炮会的负责人，一般是在村里德高望重、熟悉仪式流程的老年男性担任。由于炮会有很多是体力活，所以一般是男人参加。起炮福还会开始当年还炮节的第一次筹款活动，炮会成员首先当场认捐。此后的一年里他们还要负责到处拉赞助。不过这个过程也使得外出打工的村落成员保持频繁的日常联系，相互之间就不会那么陌生。还炮福仪式与起炮福的性质相类似，是在还炮节开始的两周前，炮会成员再一次的协商筹备会。这一次会具体商讨即将到来的还炮节的各种事宜，明确今年节日的方案与流程。还会明确还炮节当天的调度、出纳、会计、总监、各委员等职务岗位，是更细节的组织安排工作。

四、村落社会的再生产

以上是罗仁村还炮节的完整流程，可以非常清晰地看到村落社会再生产的整个过程。首先，还炮节作为一个公共性事件，吸引了全村人参加。尤其是2002年村支书JXR的节日仪式改造应和了当代社会的特点，将还炮节的仪式变成了全村人的狂欢节，尤其受到年轻人的欢迎。所以，即使今天村中大部分年轻人都在城市打工生活，但每年的还炮节都会回到家乡参加。另外，因为龙母信仰的支撑，还有那么多显灵故事流传，以及人们对美好生活的愿景，村人对还炮节的每一个活动都充满了热情，而且异常积极主动。我们发现，原本具有重要意义的祭拜祖先活动不再成为人们活动的核心。人们都自发、积极地参与到村落集体活动中，这正说明了村落社会有再生产的动力，个人能够被很好地结合到社会集体中。

其次，还炮节是村人自行组织完成的活动。其重要的组织——炮会，虽然是临时性的，但是其以炮头为中心，以中队为基础，结构是相对固定的。每个中队的成员类型也都大致相同，首先中队里熟悉节日流程的那几个老年人一定是炮会的核心，其次中队里年轻力壮、离家不远、每年会回村的就是炮会的主要人员，最后乐于捐款的、善于筹款的村人也都是炮会要网罗的对象。在这个村落社会的自组织中，不同的人有不同的作用，承担不同的责任，哪一个部分如果未能完成好，那么整个活动就会受到影响。而这个活动的内在核心是民间信仰，对龙母能够保佑添丁发财的共同认知，让炮会的每一个成员都能够认真负责，而且还会努力监督别人。这就保证了社会自组织的良好运作和效能。

再次，我们不难发现罗仁村的还炮节是有着历史基础的，但是又与历史上的节日有许多不同，这些不同刚好都体现了乡村社会公共秩序的形成和合作机制的孕育。比如，游神活动是先去祭拜梁村的土主再回来祭拜上村的土主的。另外，梁村的添丁炮是与上村的添丁炮在一起燃放的。这些都说明罗仁村通过仪式将梁村与上村紧密联系在一起。而与此同时，大乌坑这个焦姓的分支显然被排除在了外面。再比如，在放添丁炮之前的拜龙母仪式上，炮会负责人会先让为本次活动捐款的前十名上香祈福。这显然是适应现代经济社会中的一种变化。这个变化对于村人来说会刺激他们更大方、更慷慨地为村落公共事件服务，即使不在村中也能够通过捐款和村落保持紧密的联系，为村落做出贡献。以至有村民告诉我，这就是主人翁的感觉。还比如，抢添丁炮之后，放添丁炮的主人家会主动给抢到炮箍的人送上礼物或者邀请吃饭，这都让村落人保持紧密的联系，并产生互惠机制保持村落社会结构的稳定。这些都是一个社会能够良好运行的表现。

最后，我想着重谈一下改造的抢炮箍仪式。这个仪式很有特点，它在今天已经类似于一项竞技体育的比赛了。这几乎就是村中以中队为单位，以青年男性为主要成员，展开的一场有基本战术安排的比赛活动。而且特别要指出的是，这个抢炮箍的活动有明确规定，那就是外村人绝不能参加，哪怕你是本村人的姻亲或者好友，都不准许加入，所以是一个社区边界明确的集体活动。在现场看来，这个活动异常震撼和惊心动魄，而且具有一定的危险性，但是每一年村人的投入程度之高，认真程度之大，都令人难以置信。这说明，社会的个体成员，通过仪式，产生强大的激情，并且建立起认同感。而且仪式不断提醒人们，群体总是比它的任何单个成员重要（夏循祥 2014）。因而个体对村落社会产生依赖感，关系紧密，我觉得这时候仪式的重要性远远超过了信仰本身。

五、讨论

我们都知道，在现代化的背景下，中国的传统乡村社区正面临着巨大的考验。罗仁村通过仪式的改造和还炮节的庆典，顺利地调适了社区内部的关系，维持了社会再生产机制。不过，我们也同时要注意，中国社会的公共性，从长期的历史来看是权威主义与权力主义混合型的公共性，就是说中国社会中的权威很难持续性生产与再生产，从而很容易沦落为没有道德的权力（李明伍 1997），罗仁村的还炮节其实也隐性地面临这样的危机，在目前看似良好的运作机制的背后，是 JTC、JXR 等一批具有个体道德性的成员在默默付出，但今天村落年轻人所经历的成长环境和教育模式，使他们的道德操守和社区领导能力令人担忧。

注释

[1] 因罗仁行政村与自然村名称重复，所以本文后续将按照村里人的习惯将罗仁自然村称为上村，将梁村称为下村。

[2] 2013 年所见乃焦氏第九代孙焦洪君所修的族谱，2016 年焦氏族谱在村支书焦同昌的努力下已续修完成。

[3] 今年刚刚诞生了十七代的第一个成员。

[4] 梁氏族谱是梁村的梁锦贤先生提供的其自己抄录的《梁氏历代族谱后存》，但内容并不完整。

[5] 2013 年笔者第一次调查时，村中有四位 90 岁以上高龄的老人健在，除一位梁奶奶叙述混乱外，其余几位爷爷所述内容有很大的相似性，因此笔者比对几人访谈内容整理出传统时期罗仁村“二月二”的历史。

[6] 当时村中有一座龙母庙，就在今天村委办公楼的位置（见图 2），因此每年的龙母祭拜都在庙里举行。

[7] JZD 在村人的口中是非常善良和勤劳的人，在 2010 年为了保护自己的侄女被车撞死了，村人都为他惋惜。也有些老人告诉笔者，就是像他这么善良这么好的人才能够捡到龙母牌位。

[8] 牌位是铜制的，据说裹了很厚的防雨布，所以捡到时完好无损。关于这件事，笔者多方求证，村人说法几乎一致。

[9] 村人说，龙母巡游是隔了几年才开始恢复的。因为担心被说搞封建迷信活动，村里组织活动都很小心。

[10] 这里要说明一点，龙母巡游恢复时就有当地的戏班跟随。不过外请的文艺表演队是没有的，2002 年是第一次。

[11] 对于这个叫法，村中有些老人说，建国前就有这么叫的。不过，我们可以认定村中人对这个节日基本统一叫法就是从这个时候开始的。

[12] 1989 年恢复节庆以来，因为不敢大办活动，丁财炮都是统一放一次就结束的，没有各家燃放的习惯。

[13] 这里的开耕节就是二月二，不过政府要命名为开耕节，但罗仁村希望坚持自己的叫法还炮节，一度僵持不下，最后在老村长 JTC 和老支书 JXR 的调解下，村口挂出了“庆祝双二节”的标语。第二年政府的宣传结束后，罗仁村又继续着他们的“还炮节”称呼了。

[14] 村人接受外乡人的捐款，也同意外乡人参与龙母上香活动。但是不愿意外乡人抢炮箍。这是非常明确的公共规则，在访谈中村里的老老小小都会提到这个原则问题。

[15] 其中 2014 年因个人原因中断一次，但后来进行了访谈补充和当年节日相关资料的收集工作。

[16] 老人的儿子在广州开工厂，虽然不常回来帮忙，但每年过节的当天几乎都会回来。只是老人觉得他还是太草率了。

参考文献

高丙中：

《社会领域的公民互信与组织构成：提升合法性和应责力的过程》，北京：社会科学文献出版社，2016 年。

《妙峰山庙会的社会构建与文化表征》，《文化遗产》，2017 年 6 月，第 72-79 页。

李明伍：

《公共性的一般类型及其若干传统模型》，《社会学研究》，1997 年第 4 期，第 108-116 页。

彭文斌、郭建勋：

《人类学仪式研究的理论学派述论》，《民族学刊》，2010 年第 2 期，第 13-18 页。

夏循祥：

《社会组织中公共性的转型——以广东省坑尾村家族组织为例》，《思想战线》，2014 年第 6 期，第 84-89 页。

徐天基：

《村落间的仪式互助——以安国县庙会间的“讲礼”系统为例》，《宗教人类学》第四辑，2013 年，第 217-246 页。

书评

评《制造：人类学、考古学、艺术与建筑学》

余昕*

Making: Athropology, Archaeology, Art and Architecture. Tim Ingold. Routledge. 2013.

《制造：人类学，考古学，艺术与建筑》，Tim Ingold 著，劳特利奇出版社，2013 年。

阐释人类学大师格尔茨有一本名著题为 *After the Fact*，用巧妙的双关暗示了人类学的志业，在于“后事实追寻”。言下之意，人类学的知识源于事件发生之后的解释和阐释，而人类学者之于事实的关系，则是追逐和寻求。一代又一代的学者将这样的人类学知识论奉为圭臬。然而，在读完 Ingold 的 *Making: Anthropology, Archaeology, Art and Architecture*[1] 之后，关于人类学研究与世界、事实、物质、时空之间错综复杂的关系，可能会有不同的思考。

人类学者从小地方研究大问题，往往陷入对国家、民族、社会、

* 余昕，重庆大学人文社会科学高等研究院。

政治、经济等重大问题的讨论，却容易忽视一个兼具宏伟和细微的层次，即人之为人或人类存在（being）方式以及人类与物质世界的关系的讨论。在本书中，Ingold所做的正是提醒我们回归这一根本。这立即让人想起马克思关于劳动构成了人类本质的论断，此书在某种意义上也可以视为Ingold以“4A”为例对这一论断作出的具体阐释。“4A”代表的四门学科恰好都是当今高等教育的异类：它们通过制造（making）而思考（thinking），而不是相反，它们反对将理论家和实践者分别置于学术世界的内和外（p. xi）。如果我们抛开之前所有关于制造的成见，而将制造（making）与生长（growing）类比，会发现制造，“是制作者和物质的对话（correspondence）”（p. xi）[2]。

什么是知识（knowledge），怎样获得知识？或许Ingold根本避免使用“知识”一词的原意，就在于它作为一个名词，暗示了知识的具象化（embodiment）和死亡，而忽视“知道”（knowing）作为一个过程和动作之重要性。在别处，他也反对使用“传承”（transmission）来描述知识的获得。这是因为，仅仅提供信息并不意味着获得了知识（p. 1），认知意味着与认知对象的共同生长、让其成为自己的一部分。认知是主动追寻或追随的过程，知识不是外在于人的目标和对象，在认知的过程中变化的不是知识的积累，而是认知者的生长和变化。在这一主动的动作中，信息并非由某个具有能动性的主体（教师）提供，认知者获得的并非关于世界的讯息，[3]而是让自己被世界教授。

这样的区分非常关键，因为它构成了Ingold眼中人类学（Anthropology）和民族志（Ethnography）的重大区别，二者代表了完全不同的知识论。人类学的田野工作意在“与人的研究”（study with people），而民族志则意在“对人的研究”（study about people）。对于前者而言，研究者最初在研究对象的社会中如同新生儿一般无知，他/她需要逐渐学习如何看待和聆听、感受研究对象的世界。虽然这个过程并不排斥以民族志的方式记录——事实上，通常情况下人类学家兼具民族志学者的身份，但正如作者反复强调的，二者的区别和认识上的混淆正是诸多麻烦的根源。

作者总结了二者三项重大的不同（p.3）：首先，就研究者与研究对象的关系而言，人类学中二者是对话关系，类似于一次漫步，二者渐渐磨合协调到步调一致，最终研究者和研究对象构成一个沟通的整体；而在民族志中，二者的关系是面对面的主-客对立。就研究的趋势而言，人类学研究者在田野中学习，并从习得的知识、技能中推衍（像一个当地人那样）对于世界的判断；而民族志研究者在田野中调查，试图从收集的信息中归纳出趋势和规律。最后，人类学的目标在于转变（transformational）和升华，民族志的目标在于记录（documentary）和描述。

这是理想的区分，没有任何田野工作是纯粹的记录或纯粹的转变，但作出这样的区分却至关重要，因为它凸显人类学作为一个学科的特殊知识论和本体论特征。同样这也不意味着这是“理论”和“经验”、“手段”和“目的”、或“理论”和“方法”的分别。作者的论述正在于驳斥这一系列的二分，他反对那种认为可以将物（things）在真空中理论化而忽视其所处的真实世界、进而将这种理论用于解释世界的做法。相反，二者都是“技艺的实践”，对于作者而言，人类学就是研究者在与物质世界的融会、参与和感悟中，获得知识、实现生长的方式（p. 4）。

这一切的基础，在于人类就是物质世界和对象世界（如果非要这么区分）的一部分。这也是人类学引以为傲的参与观察的真义。参与观察是与主流社会科学的实证主义完全不同的人类学特有的技艺，它意味着研究者身处当地人现有的特定时空的世界之中，以对这一世界深厚的理解和领悟为基础，最终像当地人那样和当地人一起想象世界和生活应该是怎样。讽刺的是，在如今的学术知识生产模式中，由参与和实践获得的领悟成为了某种有待阐释的经验资料，而不断割裂研究者的所在与认知（p.5），参与观察成为了收集数据的方式，被剥夺了本体论的意义。

知识并非在当地世界之外、由理性的头脑构筑，而是在直接、实在、感官的对周围世界的参与过程中，获得了特定的感知和判断能力之后的所见、所听、所感。人类学者所获得的知识不是因为他站在当地世界之外，而是因为他就是当地人，人类对于物质世界的了解不是因为人站在物质世界之外，而是因为人类就是物质世界。如此，知识可以被安放回它应在之处，即存在的核心（p.6）。

在此意义上，人类学者更加类似于手工艺者，二者的知识都根植于周遭的物质世界，生长于自身与它们的交融之中。这是一种试验和探索，没有预先的计划和猜测，这种方法不在于描述或表征，而在于打开我们的感知能力，感知和应答（respond）世界。也是在这个意义上，"4A"都通过实践的方式与世界进行对话。因此，考古学的工作不是对艺术和建筑历史的研究，艺术不是艺术品（物品、光影、行为）的收藏，建筑学不是建造物（built structures）的集合。这样的误解，源于某种关于物质性（materiality）的观念。

所以我们应该反思"物质性"概念本身。当物质文化研究学者谈论所谓"物质"时，他们实指业已制作完成的器具（objects），而非组成这些器具的材质（material）。这种差异源于西方哲学对内容（matter）和形式（form）、运动（flow/movement）和静止（stop/stable）的区分，以及人类学根深蒂固的关于文化和自然的争论。想象两条永不相交的纵向延伸线，一条代表思维的流动（flow of consciousness），另一条代表物质的流动（flow of materials），我们通常所做的是横切这两条线，得到的两点，对于前者而言是象征（image），对于后者而言即是器物（objects）。二者的横向关系，构成了词语与物质之间的关系。很多时候，“4A”的研究都集中于横向的这条线，即探讨图像与器物之间的静止的相互关系。在面对野蛮的、荒凉的物质世界时，人类只能将自身抽离于这一世界，然后重新将自身安置于另一个象征的、语言的、文化的、心智的层面，才能对它以及自身进行理解（p.27）。作者在本书中所反思的正是这样的物质性，其方式是将视线从横向变为纵向，从静止变为运动，从视角中心变为感觉中心，探讨图像和器物形成和不断流动的过程及这一过程所涉及的感官。

这一过程即是 Ingold 所言的制造（making）。我们习惯将制造视为一个工程，即从一个头脑中的设想或蓝图开始，结合一定的原材料，将这一意图实现的过程。其结果即人工制品，对它们的研究则成为了物质文化研究。这种将观念经过一定组合实现为物质的过程，源于古希腊的形质论（hylomorphism）（p.22）。而其问题在于假设了一个固定的形式和一个无差别、同质的内容（matter）。与此相反，现实世界中的物质是流动和变化的，因此也是可以追踪（follow）的，手工艺者和实践者，实际上是漫游者和旅者，他们所做的是参与到世界的形成过程中，将它往可欲的方向引导（p.25）。这个过程类似于打铁，铁匠需要不停地将铁融化重铸，以调整形状，只有在冷却之时，铁器的形状才得以固定，因此铁器实际上是在铁匠的击打、熔化、塑形、再次塑形的过程中成型的。任何器物都处于类似的人和物不断融合、相互形塑的过程之中。

因此作者提议，将人类从以自我为中心的神坛上拉下，更为谦逊地对待自己周遭的鲜活之物，将制造视为“生长”（growing）的过程。制造不是制作者的垄断，而是他与物的合力 / 合作（join forces），物也不再是被动地接受宰制之物，而具有自身的特质和力量，能够参与制作的过程。在这样的视角下，一块大理石的风化与雕刻家制作的石像之间没有界限，只不过在大理石的生长过程的某一阶段，雕刻家带着他的石斧参与了这一过程。同样的道理也适用于建筑。通常认为的建造，大部分的创造性工作都集中在设计环节，随后的修建不过是设计的执行和实现，设计实现之时，亦是建筑完成之日。然而这不过是一个幻觉，在所谓的建造完成之后，建筑物的生命仍未完结，雨水冲刷、阳光暴晒、蚊虫滋长，这些被视为“破坏”建筑物的过程，不过是其生命历程中的一部分。一个“自然”形成的土堆和一座“人为”建造的纪念碑，二者的区别似乎并没有想象中那么巨大。

我们通常将建筑和纪念碑的集合作为历史的印记，但这样做的同时，历史的生命力也被终结。所以纪念碑的吊诡即在于，它们建造的成功，代表了其建造目标的失败（p.78）——正是由于它们的存在，本应纪念的历史终结了，过去成为了死物，与现在和将来无关，像一条搁浅的鲸鱼，被困于沙滩，而时间继续流逝（p.79）。因此，纪念碑的存在使失却的过去和鲜活的现在之间的距离渐行渐远，或者它们最多成为了个人记忆的组成部分，而与纪念的初衷本身无关。“过去”对于“当下”而言，成为了另一个世界。但我们会对山川提出同样的问题吗？如果说山川的非人造特征让类似的问题显得荒谬，那所谓“古迹”又在多大程度上是人造的？古迹和山川的分别，即“自然”和“人造”的对比，以及“古迹年代历史”的探究，源于形质论视野中关于“制作”的观念（p.81），即将纯粹的抽象形式加诸纯粹原始 / 天然的材料，将它变成人工制品。

在这里作者实际上暗示了另一种物质与时间的关系。历史学和考古学之所以重视古迹和器物，是因为它们对于记录（record）时间的意义，它们在时间流动之外，是结绳记事式的结点的代表，但作者提出，如果将它们重新放回流动的时间呢？那我们和过去与未来的关系都将不同。

作者用讲述（tell）来表明这种状态，他认为讲述至少有两层含义：第一，表明一个人知道关于世界的故事；第二，表明一个人能够觉察周围环境的微妙信息，并对此作出判断（p. 110）。猎人就是最佳的例子，他们往往腹中有很多故事，也能准确地判断和预测猎物的行踪。这两层含义紧密相关，因为“讲故事”亦是训练新手的方式，正如人类学家通过听故事逐渐融入当地世界，他亦在同一过程中获得了和当地人一样的判断能力。因此，“讲述”并非是具体信息的提供或指令的传达，而是引领，由此让新手与先辈处于同一世界之中。同样，“讲述”也并非辨析（articulation），即思想的拼凑和组织（joined up thinking）（pp. 110-111），因为说出的故事传达的并非具象本身（specification），而是通过具象方式进行的引领。辨析假设了预先存于脑中的想法，它拒斥知觉和同感，是理性之友，感觉之敌（p. 111）。如作者为全书主旨所绘制的两条并行线，辨析代表了横向的词与物、形式和内容之间的关系，而非纵向的互相交织的运动。讲述获得的知识源于实践者与物质的交汇中生产出的知觉。

接下来的问题是，如果讲述与知觉相关，那执行这一动作的器官是什么？人体头部的眼、耳、鼻，显然都具有感知能力，却不具备表达能力（虽然眼多少算个例外），喉具有表达能力，却不具备感知能力。作者认为，只有手才是人类区别于其他所有生物之外，即真正的人性所在。它们不仅具有触觉，还能通过手势、书写、绘画和编织进行表达。并且，手并非大脑的仆从，而毋宁说是大脑的延伸（p. 112）。在此更为重要的是，作者并非是在器官或解剖学的意义上称手构成了人性所在，“人之手之所以代表人性的原因，并非在于它是什么，而是由于它做了什么”（p. 115）。换言之，人性不在于作为器官的手，而是这双手具有的类似于身体技术的技艺和能力，它们来自于过去实践的历史。在双手的身体技术中，人类、工具和物质构成了一个器物得以形成的整体。因此，在知觉和制造（书写、绘画、手势）的双重意义上，双手能够进行讲述（p.116）。

用手讲述，表明人类并不处于事实之外（民族志）、之前（设计、预言）、之后（阐释），而研究人类世界的人类学是一门“手艺”。讲述，也是 Ingold 的写作手法，或许因为本书是一门课程的过程的结晶，阅读此书就像一次思维的漫步，没有预先的设计，在与作者的同行之中、引领之下，不停地发现、提出和解答问题，同时留下更多的问题和启发，继续前进。

注释

[1] 本书的副标题为“4A”，即 Anthropology, Archeology, Art 和 Architecture, 这是 Ingold 于 2004 年在 Aberdeen 开设的一门课程的名称，这门课程的收获，最终构成了本书的主要内容。

[2] 将 correspondence 翻译为对话源于 Ingold 自己所举的例子，即书信的来往，原文为，making is a correspondence between maker and material, and that this is the case as much in anthropology and archeology as it is in art and architecture（xi）。

[3] facts *about* the world，斜体为原文所加。

评《邻家政府：中国城市中的社区政治》

何晴*

The Government Next Door: Neighborhood Politics in Urban China. By Luigi Tomba. Ithaca; London: Cornell University Press, 2014.

《邻家政府：中国城市中的社区政治》，Luigi Tomba 著，伊萨卡和伦敦：康奈尔大学出版社，2014 年。

2002 年，Luigi Tomba 来到中国，最初目的是调查中国新兴的中产阶级的日常生活和政治，但一些不符合他最初预期的现象吸引了他的注意。他此前以为住在北京新兴的封闭小区内的大多是受益于改革开放的企业家，但很快发现那些高档小区的居民大部分是在体制内工作的白领或者专业人士。中国社会阶层改变的途径也出乎他意外——个人和家庭的阶层上升直接通过房产，间接地通过原生家庭的社会地位，和收入关系不大。Tomba 在 2015 年接受《纽约时报》采访时提到，他认为中国兴起的这个阶层不是一个真正的新中产阶级（a new middle class），而是一个新的有房产的阶级（a new property-owning class），

* 何晴，威斯康星大学人类学系。

因为这个阶层只是在财富上很新，但在政治上并不新——依旧与政权保持长期关系，并依赖这种关系获取财富。或许因为这个原因，他意识到小区是研究中产阶级如何发展自身身份的理想地点。

尽管 Tomba 无意于把小区看作是“政权-社会关系”的隐喻，但他认为：“居住区的社会和空间面貌包含并再现了特定的权力关系，界定了对社会分类很关键的分散的空间类别，重新铸造了身份，创造了社会关系，限定了新经济利益的界限，并助长或包容了冲突”（pp.3-4）。Luigi Tomba 选择把两个居民小区作为研究对象，一个是位于沈阳的工人阶层老公房小区，一个是北京新兴的中产阶层封闭小区。他认为中国的居民小区中存在许多政府管理，因而研究居民小区中的权力的日常行使，可以“揭示利益日趋复杂化的个人、家庭和群体与国家目标互动的各种方式”（p.4）。

很多西方媒体持有印象，中国社会的日常生活中必定随处可见民众和政权的冲突，比如新兴的中产阶级会越来越多地要求参与政治决策，比如被剥夺至赤贫的农民工和下岗职工会挑战政府统治，造成社会动荡。Luigi Tomba 却发现事实并非如此。他注意到：各方在冲突中不会提出一些革命性的口号，相反，他们会试图把主流政治话语作为维护自身利益的话语框架。譬如在北京的商品房小区里，居民在面对开发商维权的时候，用的理论框架不是“公民社会”初期的人们经常会用的那些，而是“道德”、“国家建设”、“爱国主义”、“素质”、“对现代化贡献”等话语。在沈阳老公房里居住的下岗职工会在冲突中不断提及中央政策的社会主义精神，以此指责当地政府没有给他们提供足够的社会福利和援助。政府会用“和谐”、“素质”、“安全”这样的词语在小区中推行自己的政策，而开发商也会用同样的词试图把高档住宅卖给那些渴望提升阶层的消费者。Luigi Tomba 提到，无论这样的表达是一种应对策略还是一种结构性的文化霸权（cultural hegemony），它的实际效果都是使政府行使于日常生活中的权力合法化了。当然，他也注意到这个过程是两面性的，一方面，市民们内化吸收并接受了政府行为，另一方面，他们逐步发展出自己的能动性（agency），并以此提醒政府别忘了把空洞的话语落实为具体的行动。

The Government Next Door 一书中最重要的概念之一是“社区共识”（neighborhood consensus）。Luigi Tomba 对这个概念也有详细而清晰的说明：尽管共识并不代表政治支持，但是它的最终底线是不改变社会制度以及维持社会秩序；这种共识是允许争议的，甚至经常是在冲突中才体现出来；共识本身是居民和不同级别的政府讨价还价的筹码，可能会影响长期的权力关系；中国不是一个共识社会。

那么政府是如何渗透进小区这个层次，并与其居民达成了共识的？在 Luigi Tomba 看来，这种社会与政权之间的共识并不是强权政治强加于民的，而是通过无数次的协商和争执构建的。政权如果要被接受，并证明其合理性，其行为必须被市民内化吸收并认可，于是社区层面的管理就显得尤其重要，因为社区管理可以帮助维持社会稳定并创造市民对政权的依赖和忠诚。基层的政策落实可以影响人们对政权统治能力的感知。Luigi Tomba 引用了米歇尔·福柯的观点：权力如果只是一味地压制和说“不”，人们是不会遵守的……权力必定“同时也讨论和生产事物，引导快乐，形成知识，创造话语”。（p.17）

Luigi Tomba 提到，随着城市人口越来越复杂，城市空间越来越隔离，其实也是在满足政府对空间分类、维持社会秩序的需求。他和一些学者一样，把房地产市场的放开看作政府对市民社会干预的加强——以前政府是通过单位大院，现在则是通过不同等级的私有空间，如封闭小区来对空间分类，并通过居委会、街道等早已存在的结构对私有空间进行管理。日常的民居-政权接触，也体现出政府统治在今天的中国越来越灵活。政府允许人们在一定私有空间范围内通过自治组织，即业主委员会进行自治，并由私有的物业管理公司提供相应服务。这么做可以使政府肩上的中产阶级的负担变轻，而把更多资源放在需要帮助的人群上。这样的策略被 Tomba 称为“微管理”，和经济开放政策是一体的，毕竟政党在国际和国内的统治合理性很大程度上建立在其经济是否成功的基础上。

Luigi Tomba 认为社区层面的政府管理有五种行为：社会聚集（social clustering）、微管理（micro-governing）、社会设计（social engineering）、被包容的斗争（contained contention）以及榜样论（exemplarism）。本书也是按照这五个标题写成了五个章节。

在第一章“社会聚集”中，他提到房产私有化让人群分类地聚集在不同空间里，从而导致政府采取更灵活的管理方式。但社区服务一直没有真正进入那些依赖物业公司提供服务的高档小区，而主要集中在穷人和残破的小区，因此那些富人小区通常比工人阶层小区更远离政府。

在第二章“微管理都市危机”中，他则研究了在沈阳的工人阶级小区中，那些基层的政府机构（如社区委员会）是如何为大量失业居民提供服务的。他还特别注意了那些基层党员在这些社区中所扮演的角色，他们一方面复制官方的话语，另一方面他们自己也是改制的受害者，替那些边缘化群体发声。

在第三章“房屋和社会设计”中，Luigi Tomba 提到有房产的中产阶级的增长其实是社会设计的结果，政府的房屋补贴政策以及对居住空间隔离的支持都向那些体制内的群体倾斜。

第四章“被包容的斗争”则关注在小区内日益多见的冲突，但 Luigi Tomba 也注意到冲突只要发生在封闭小区的围墙内，便不会引起政府的关注，也不会威胁政权稳定，引起系统动荡。

第五章，那些住在新小区里的中产阶级成了社会的榜样，他们自力更生，遵纪守法，高素质。城市复兴和改造传统的城市中心为后工业时代的全球化大都市，首先就需要依赖宣传中产阶级的榜样论，在这个过程中产生的“价值观既是金钱上的，也是政治上的”（p. 25）。

得益于卓越的洞察力，Luigi Tomba 非常公允和客观地评价了中国社会现状，清晰阐述了中国极为复杂和微妙的社会-政权关系，是近年来描写中国社会最深刻的著作之一。本书揭示出中国政府的房产私有化和此后一系列租房住房补贴政策在创造财富和社会地位中发挥的作用。无论政府政策初衷如何，这些政策实际加剧了地区之间、城市和农村人口之间、有社会关系和无社会关系的家庭之间的财富不平等。我在上海居民小区做田野调查时的发现也印证了 Luigi Tomba 的观点。房产，而不是工资收入，成为居民提升阶层、财富增长的最重要手段。中国有房产阶层的绝大部分诉求的核心是房产价值。小区内发生的维权只局限在居民和开发商、以及和物业管理公司之间。冲突双方都试图用官方话语来合理化自己的诉求和立场，并试图以此来赢得居委会和街道等基层政府机构的协调。

The Government Next Door 一书贡献了相当深刻的见解，但在一定程度上也不可避免地受到时间和地点的限制。2017 至 2018 年，我在上海某个中产阶级小区做田野调查。据我的观察，业主并非大多是体制内的人和专业人士，而是有私营业主、外企员工等各式各样的职员。首先，这可能是因为上海和北京这样的政治中心有着不同的政治环境，中产阶级的职业构成本身不一样。尽管上海业主在维权时也经常会使用一些官方话语来作“护身符”，但其程度比北京相对弱。Luigi Tomba 也提到，每个城市的背景不同，需要用不同的方法去研究。其次，Luigi Tomba 的田野研究集中在 2002 年至 2006 年间。近年来在上海等都市中，商品房封闭小区已成为居住空间的主流，其居民自然是非常多样化的，居委会等政府机构与居民的日常交集也趋于极小化。

评《爱的不确定性：当代中国育儿的政治与伦理》

吴江江*

Love's Uncertainty: The Politics and Ethics of Child Rearing in Contemporary China. Teresa Kuan. University of California Press. 2015.

《爱的不确定性》，关宜馨著，加州大学出版社，2015 年。

关宜馨（Teresa Kuan）教授《爱的不确定性》一书所关注的育儿问题切合了时下许多中国中产家长共同的焦虑。该书文笔优美、理论丰富，作者的访谈及参与观察涉及众多对象，包括心理健康咨询师、学校老师、中产家长等，同时对流行文化文本（育儿手册、电视剧等）和国家政策亦有探讨。关宜馨以消费能力（consumption power）为标准对中产这一群体进行了定义（p.17），并通过细致的分析，展现了中国中产家长，尤其是母亲，在充满矛盾的教育理念和竞争激烈的社会环境中既难逃焦虑，又竭力为子女创造“条件”的过程。她向读者展示了中产父母们在此过程中面临的育儿困境：在国家政策、流行教育

* 吴江江，威斯康星大学人类学系。

理念和社会现实的共同影响下，他们一方面希望培养孩子在严酷竞争中脱颖而出的技能，一方面又希望他们发挥个性，拥有自由。关宜馨将这种情境下父母的应对之策称为“安排的艺术”（art of disposition），并指出这种育儿方式是一种伦理实践，且背后有中国传统哲学思想作为支撑。

关宜馨在书中解释，“安排的艺术”（art of disposition）一词源自于福柯1978年关于“治理术”（governmentality）的演讲。福柯指出早年欧洲认为治理的艺术在于对事物正确的安排，以带来便利的结果。基于此，她将掌握这种“艺术”的人定义为“既能认识到人所处的具体情境的影响和约束，又能知道何时何地改用技巧操控、安排”（p. 21）。她认为，所谓“安排的艺术”一直贯彻在中国人的哲学思想中，这种观点认为治理的艺术不来自于强力，而在于对事物的安排，通过安排、创造情境和条件来达到期望的结果。从而，要达到某种治理效果，安排、创造情境比了解个体的特点、个性更重要。她以《孙子兵法》、“草船借箭”等古典文本和故事为例，指出在中国传统思想中，胜利来自于为个体找到最合适的安排，并在正确的时机采取行动。同时，不同于福柯笔下的治理术，在中国传统思想中，“安排的艺术”虽然也强调环境因素比主体因素更有决定性作用，但还是给个体的操控、安排留下了余地。比如，中国文化对“势”的强调就是这种“安排艺术”的最好体现。

本书强调，这种“安排的艺术”同样体现在当今中产家长的育儿实践中。中国中产父母深知把握“势”的重要性。然而不同的是，当今中产父母的“安排”带有明显的伦理色彩。父母如何以及何时调用个人能动性（agency）以帮助孩子应对现实的竞争、障碍和约束，不仅和中国的政治环境相关，同时更是一个重要的伦理问题：“当今中国，育儿问题不仅是政治更是伦理议题”（p.7）。书的第四章——“创造‘条件’或‘安排的艺术’”——就以具体的例子和民族志直接讨论了这个问题。作者指出，热衷于为孩子“创造‘条件’”的种种想法和做法都体现着中产家长们“安排的艺术”。在充满竞争和不确定性的社会环境中，孩子的未来发展由物质、经济、生物、营养、社会、历史等各种“条件”所决定，父母，尤其是母亲在此过程中就成了“创造‘条件’”的伦理主体。行动或不行动，在何时行动，都迫使家长不断地做出选择，因为这是他们唯一能够为孩子确保的事情。这种道德责任背后充满了对倘若自己不作为、未来可能出现何种后果的担忧。

然而，不同于西方（尤其是美国）中产家长，中国家长面临的一个特殊困境在于，社会中存在着各种互相矛盾的教育理念。对中国家长来说，教育不仅仅意味着尽力而为“创造‘条件’”，尽力是不够的。家长到底该怎么努力？往哪个方向努力？作者在她的被访人中发现了诸种互相矛盾的教育理念。一方面，国家大力推动“素质教育”，旨在提高人口质量，但对“素质教育”的定义极为模糊，官方话语和各路教育专家都有不同的解读，对它的理解混杂着所谓的西方教育理念，主张关注孩子的个性，给他们充分的自主性。而另一方面，严苛的考试体制又与前述的教育目标大相违背，迫使家长不得不竭力培养孩子的竞争力，而不是任凭他们自然地发展。

在这种境况下，关宜馨在中国中产家长的育儿实践中读出了一种“尝试的伦理”（ethics of trying），它和“安排的艺术”相辅相成，形成了当今中国中产家长的伦理主体性。她指出，家长的希望（hope）里蕴藏着面向自己的道德压力和伦理责任。对孩子寄予希望即意味着在特定的情境下做出反应，比如何时放任、何时严厉，抓住每一秒为孩子创造“条件”、提高竞争力。因此，所谓“尝试的伦理”，“不在于遵循标准化的道德准则，而更多是一种实践的哲学，它将因果、效率和责任视为最主要的考虑因素”（p. 18）。它也给父母提供了一种道德慰藉：在矛盾的教育理念和在高度竞争的社会中，我或许没有办法保障你的未来，但我至少在关键时刻尽可能尝试了。

由此，作者挑战了传统马克思主义关于意识形态的定义。经典马克思主义将家长的育儿行为置于资本主义制度之内进行分析，认为家长，尤其是母亲，在不自觉中接受了现存的意识形态，为资本主义市场不断生产劳动力。关宜馨却认为，中国中产家长的例子说明了他们对自己所面临的社会环境和生产关系是有深刻意识的，也知道他们的孩子有怎样的特权。他们绝非无意识地受国家意识形态和市场的愚弄，他们并非不知道国家为提高人口质量而宣传的"素质"话语，也不是不知道经济自由化带来的教育商品化、市场化的问题——"他们知道，但他们别无选择"（p.163）。他们在对所处社会环境充分理解的基础上，尽可能地调动个人能动性，努力应对育儿困境。作者进而希望通过这样的分析重建一种结构和个人、或者说宏观和微观之间的联系，她希望通过中产家长的育儿经历回应一个问题：当代中国社会有许多自上而下的权力和限制，这种情况下我们该如何描述个人能动性，避免将活生生的个人经验化约到宏观历史进程之中？如前所述，"安排的艺术"和"尝试的伦理"即是回答该问题的一种尝试，作者在过于强调外部环境决定性和过分强调个人能动性之间找到了一种平衡。

然而，本书存在的一个问题恐怕在于它对传统文化资源的挖掘和解读。纵观全书，作者始终试图在中国传统哲学中找到当下家长育儿思路的根源。除了上文提到过的《孙子兵法》、《三国演义》等作品，作者还提到了《商君书》、《论语》、《孟子》等古典文本和王夫之等历史人物。然而，概括古典文本并将其与当代对接的做法似有将"中国文化"本质化的嫌疑，从而呈现出一幅稍嫌静态的"文化"图景。尽管作者并非没有注意到这点，"我无意通过梳理儒家伦理展示一个不变的中国文化内核"（p.101）。但是，将当代社会现象回溯至古典文本并非易事，作者选择的文本来自不同时代，它们所展现出来的思想内涵在什么程度上可以互相联系，又在什么程度上需要知人论世？这恐怕是《爱的不确定性》一书没能很好回答的。同时，这又指向本书的另一个问题，对古典文本的诠释基本都来自学者的二手材料，比如对“势”的解读就多来自法国哲学家 François Jullien 的 *The Propensity of Tings: Toward a History of Efficacy in China* 一书。由于对古代典籍以及传统哲学的释义并非历朝历代一以贯之，如何概括提炼出一个可以与当下对接的哲学传统，恐怕需要更多的思考和追问。而这些反思在《爱的不确定性》一书中是缺乏的。

总体而言，作为一部教育人类学民族志，《爱的不确定性》对中国中产家长的解读视角独特、例证丰富、分析细致，语言也极为平实易懂。该书不仅适合关注教育、心理、道德人类学等领域的学者，也适合任何关心当代中国教育、伦理问题的普通读者。

评《从村到城：一个中国县级市的社会转型》

王博*

From Village to City: Social Transformation in a Chinese County Seat. Andrew B. Kipnis, Oakland, CA: University of California Press. 2016.

《从村到城：一个中国县级市的社会转型》，任柯安著，加州大学出版社，2016年。

任柯安用“重组”（recombinant）一词描绘中国当前的城市化进程，它的涵义是多重的：是资源与发展在全球范围的重组，是中国城乡的重组，是中国一线二线三线城市直到四线县级市版图的重组，也是20世纪90年代以来中国急剧城市化进程中社会生活方方面面的重组。首先，任柯安认为当下的城市人类学研究固然成果斐然，对城市边缘群体和结构不平等的研究也很深入，但在批判市场标准化和全球资本主义的同时，却避而不谈城市化和现代化理论。换句话说，破多于立。任柯安重审颇具影响的马克思主义发展学者詹姆斯·弗格森

* 王博，瑞士洛桑大学（University of Lausanne）社会科学研究院（Institute of Social Sciences）。

（James Ferguson），后者1999年的名著以赞比亚铜矿区的贫穷严厉批判了现代化对于发展中国家的灾难性后果，并以此为据批判现代化理论过分单一，掩盖了剥削和不平等。任柯安指出该论点过于倚重赞比亚70年代中期铜矿经济一蹶不振的状况，夸大了社会衰退和萧条，并将之放大到全球的范围。任柯安在山东省邹平县级市的田野研究表明，城市化进程并非单纯的衰退模式（model of decline）能够解释，邹平市的案例乃至中国自90年代以来的城市化案例，足以表明在衰退之外也有增长，体现在工业化、购买力、基础设施、教育和其他方面（pp.9–10）。并且，弗格森的理论是以过去欧美现代化的历史为出发点，寻找替代理论，即衰退模式；而东亚的压缩式的快速现代化进程，则要求我们考察当下的历史与情境，分析其与现代性的关系（p.9）。

在书中，任柯安强调研究县级市在中国城市化领域的意义。县级市很普通，却极富包蕴性：既包含了国际大都市普遍的消费主义，又包含了第三世界城市常见的贫民窟困境。当然，邹平市很难称得上国际大都市，但它的崛起却离不开跨国经济，甚至，邹平的崛起也意味着世界其他某些地方的衰落。具体而言，邹平市的机遇和挑战包括：计划经济时代沿用下来的每五年的城市规划，使得道路、公园和居住区有一定条理；户口制度的松绑促成了流动人口，直接推进了经济发展的规模和速度；村集体的保留，让村民“离土不离乡”，在城市化浪潮中仍保持紧密的社会关系网络，提供了部分社会福利和保障。这些机遇首先在珠江三角洲的经济特区出现，迅速发展到其他区域。任柯安认为，在邹平这样的四线城市，比起早期崛起的珠三角的大都市甚至有些优势。比如邹平对外来务工人员的户口限制不那么严格，从而防止了对外来人口过分的劳务剥削（p.20）。

书的第一部分考察邹平自市场化以来的社会转型，分析的重点分章节依次设定为规划、生产、消费到现代性的魅影（phantasmagoria，本雅明的词汇，原意为模糊的亦真亦幻的景象，此处指现代性带来的现实与虚拟的模糊性）。选取的这几个重点都加上了“重组”为前置形容词，以突出社会转型过程中的多个角色，比如国家与地方，企业家与村民，县城厂家与外资和外来人才，等等，也突出了多种社会组织方式应对市场转型的变化，并非去旧迎新，而是对历史的制度形态（比如五年规划、单位、户口）和新的机遇（比如加入世贸、包产到户、引进技术、全球市场）进行重新组合的社会过程。所谓重组，新旧的关系并非交替，而是以新的方式混合。比如计划经济时代沿用下来的集中的城市规划制度，在邹平县的案例中，对当下城市的公共产品（比如道路和绿化）起到了积极的作用，避免了拥挤或者卫生差的状况（p.23）。同时，90年代的基础设施建设已完成了县城的道路网。高速路连接到青岛港，进而直达日韩市场；高速路也通往济南机场，进而直航至世界各个市场。

2000年同济大学与邹平市政府合作，一手操办了旧城、新城和开发区的规划，三者分别承担商业、行政与居住、工业三大功能。几乎唯一的难题就出在城中村，因为城中村与周围规划的城区有显著区别，主要问题集中在安全性、卫生性和合法性，是城市规划的“灾难”（p.55）。因房租低廉，城中村是新涌入人口选择居住的区域。由于城中村的过渡状态和灰色合法性，村民总是等待拆迁赔偿，对基础住房投入也极为节约，一方面有利于城市新来人口立足，另一方面无法更新设施。总之，任柯安认为，一方面国家主导的规划有利，另一方面城中村的无规划却是必要的，因为它促成了廉价房，满足了底层的需求。在国家的角色上，任柯安持与詹姆斯·斯科特相反的观点，后者认为国家的干预总是负面的。任柯安提供了较为细致的观点，即制定和执行规划的官僚体制遇到实际的城市化过程，面对城中村违反规划的行为，持容忍和策略性接受（tactical acceptance）的态度（p.61）。依据情况，地方政府往往会针对城中村的家庭采取多变的协商策略，以不同的价格赔偿，奖励提早接受失地赔偿的家庭，惩罚坚持到后来的家庭。这种差异对待的策略往往是放在增长的逻辑下，即只要有利于国家和地方经济增长，牺牲是必要的。

接下来任柯安考察了生产，以当地两家龙头纺织企业为例，考察转型后企业如何利用开放的经济政策实现飞跃性增长。有趣的是，任柯安观察到这些强大的工厂提供给合同制工人低于市场价的住房和其他福利，相当于在某种程度上提供了与改革开放前时代类似的社会保障，他称其为“新单位制度”。当然，这种社会保障是否完全可靠，往往取决于工人对企业的信任程度。虽然企业以市场价三分之一的价格将房子卖给合同工，承诺在工人退休后原价买回，但是许多外地工人不愿意作出这样大的承诺，他们一方面不信任企业会如约买回房子，另一方面也不信任企业会一直像现在这样成功，他们宁愿自己完全承担风险，也不选择这样的住房保障。

任柯安继续考察了消费，他观察到，即便2012年国家的反腐败运动严格禁止了公费宴请，铺张性消费仍然得以继续和稳定，是政商结合的重要形式（p.110）。他也观察到，农民工群体不再只是寄钱回老家的群体，也是消费主体，比如父母花钱培训提高孩子的“素质”，雇用婚庆公司为成人的子女操办婚礼等等，又比如对奢侈品牌的讲究也逐渐成为消费市场的主流。第一部分的最后一章考察了邹平街道的景观和背后的生产方式，描绘了充满性欲和男性视角的悍马豪车广告牌，充斥在大街小巷的中国各地方和外国的菜品，独立于农村生活方式之外的农家乐广告招牌和项目，对邹平的范公故里的建筑和对其他历史的遗忘，以及流行于社会各部门的“唱红歌”活动。社会学家齐美尔（Simmel 1971）指出城市的社会关系强度会造成人的感官过度受刺激，而带来一种破坏性的和负面的作用。任柯安认为邹平的民族志例子却表明热闹可以是正面的（p. 136），是城市的另外一种社会性。

相对于第一部分较为理论的讨论，第二部分更强调展示多样和生动的事例。任柯安分析了邹平各种角色的人，以阶级和城市化参与度高低为标准，划分为当地已婚蓝领，外地已婚蓝领，城中村村民，城市中产，年轻人五大类。邹平的人口从2000年以来增长了七倍，如此迅速的膨胀让捕捉形形色色的生活变得几乎不可能。任柯安关注的主要是经济收入和消费，人们在城市化的各种住房和福利政策中的位置，人们与家庭和其他社会关系的疏与亲，力图通过具体的事例来回答，在迅速城市化的邹平，人们如何利用自身和家庭来获得政治、经济和文化资源。

针对每类人，任柯安辨析了影响人们适应迅速城市化策略的关键因素。比如对于当地已婚蓝领，亲属关系是最关键的，因为年轻夫妻依赖父母对孙辈的照顾。虽然从夫居似乎把重心偏向丈夫一边，但在任柯安看来，并不一定会对女性造成劣势（p.148）。这是因为夫妻双方父母都参与了照顾孙辈的任务，也因为女性作为母亲更关注与子女的关系，能从丈夫的相对复杂的家庭和社会关系中解放出来。相较而言，外地已婚蓝领很少能享受到父母的帮助，也很少有经济能力把父母接过来常住，这种相对分裂的家庭代际互助，也促成了他们对本地的认同度不高，难以作出深度承诺，也不大会参与公司的保障住房计划。城中村的亲属关系尤为重要，因为亲属关系往往决定个人是否有权参与村集体的分红和利益，从夫居和父系的重要性就凸显出来了。总体而言，传统的家庭和亲属关系仍然在人们的经济生活里举足轻重，但是钱的作用也不容忽视。针对每个群体，任柯安总能找到反例，即事业成功的人往往不依靠家属关系，甚至再婚、离异、分居等等，也能够享受经济高速发展带来的资源。对比第一部分相对系统化的分析，第二部分的描述性大于分析性，对案例的描绘也相对完整，以家庭为单位，叙述了每个案例的基本情况。

任柯安的这本书是不可多得的关注中国社会转型的民族志作品，由于作者数十年在山东邹平市进行深入的田野调查，对细节的把握非常精细，值得城市人类学、城市化研究和当代中国研究的学者和学生阅读。尤其是他提出将人类学置于城市化和社会转型中加以讨论这一中心议题，非常值得重视，毕竟人类学往往关注较小规模的社区，以细节和特殊性见长，却不擅长宏观的把握，任柯安试图做出以中国县级市的社会转型民族志为基础的人类学中层理论，既不同于大而无当的国别论，又不同于小型社区的特殊论。当然，这本书欠缺的部分也在于，很难说它非常成功地达到了这样的目标。诚然，我们读到了城市化过程中不同人群的不同角色、地位、动机和身份认同，但是并没有理解"重组"的是什么，为什么叫重组？另外，重组本身的含义暗示是对已有制度和价值的重新组合，这便忽略了新的制度和价值可能来源于新的文化资源。过于保守的分析，有时候也限于琐碎的细节重现，读者往往并不理解为什么会有将形形色色的家庭归为蓝领已婚这样比较不常见的分类范畴。但是瑕不掩瑜，这本书的价值在于提出了问题，给出了较为翔实的民族志材料，可以预期对将来的中国城市化和社会转型的中层研究会产生很大的影响。

评《当代中国兴起的机构养老：两个世代，一个决定》

陈怀萱[*]

Evolving Eldercare in Contemporary China: Two Generations, One Decision. Lin Chen. Palgrave Macmillan. 2016.

《当代中国兴起的机构养老：两个世代，一个决定》，陈琳著，帕尔格雷夫·麦克米兰出版社，2016 年。

随着当代社会现代化、都市化与市场化的发展，对于即将步入老龄或超老龄社会的国家来说，老年人口长期照顾已经成为迫切需要关注、无法逃避的课题。人类学家长久以来就十分关注不同文化脉络下，身处不同生命历程阶段的人与族群如何形塑一个社会对于老年的观念与想象。

人类学领域对老年学研究发展的重要贡献之一，即在于将老年视为一种社会互动下的文化情境，进而挖掘个人与社会如何产生各种有关身体、空间、精神与心理、社会经济与家庭关系的意义。老化的过

* 陈怀萱，人类学博士，现任职于台湾清华大学。

程除了涵盖生理、身体与心智的变化过程，同时也体现为社会政治、经济、亲属、宗教等脉络的交互作用，以及资源分配与社会网络的实践与调适过程。尽管人类学领域对于老年的民族志研究为老年学的发展提供了系统性、整体性的视野，但早期研究多聚焦在医疗公共卫生系统下较为正式的、专业的老年照护关系方面。随着医疗的进步，人类寿命的延长，社会高龄人口的增加，人类学家也开始关注牵连日常生活关系的老年长期照顾，探究其中复杂而多元的社会变迁过程。

照顾关系乃是通过提供照顾者与接受照顾者双方在家庭与个人为基础的日常生活实践互动的过程来赋予意义，除了体现个人家庭与社会脉络下情感与道德的价值，更能映照出社会道德与政治经济的交互作用。因此，不同社会文化脉络下的照顾议题就成为挖掘身处老龄化社会的人们在日常实践中窥看国家力量、亲属与性别关系、社会阶级、地方空间等作用与发展脉络的重要视角。

在中国，传统儒家文化对于老年阶段的价值观强调尊重长辈与孝道的实践，使得家庭一直都是探讨老年照顾议题的关键角色。20世纪90年代后期，更有政策法规明文规定：子女有赡养扶助父母的义务并安排住所。父母与子女一起居住成为多数家庭承担老年长期照顾责任的模式。“由子女来扶养父母”的居家养老照顾模式不只映照出中国传统社会价值观的日常实践，同时也体现出一种稳定的社会结构。然而，数十年来快速转型的经济、都市化、住房改革以及低生育率等变化，对这种以家庭为支持的老年照顾运作实践方式造成了很大的压力。

正值中国老年照顾文化价值观面临冲击之际，2016年出版的《当代中国兴起的机构养老：两个世代，一个决定》，非常及时、有价值。上海复旦大学社会工作学系助理教授陈琳，在本书中以家庭两代选择老年照顾方式的文化观念和决策过程作为分析视角，勾勒出中国都市家庭面对养老选择时，如何调适并且寻求传统老年照顾价值观与两代生活现实差异的平衡点。

作者选择了上海的一家公营养老机构作为她的研究场域，针对十二个家庭进行超过六个月的访谈与观察。除了了解长者与其子女的健康状况、家庭亲属关系、居住条件，也挖掘他们对于过去家庭照顾与现在机构照顾的态度转变与调适。

上海是中国最早步入老龄化的特大型城市，老年人口比例增长明显，远高于全国平均水平。尽管多数老年人的养老方式仍以自己照顾自己或依靠家人照顾为主，但随着上海老龄化与空巢化现象日益严重，机构养老模式也开始渐渐成为老年照顾模式的选项之一。

本书除了导论之外，共分为 8 个章节。第二章首先描绘了养老院与中国都市区老年人长期照顾发展的社会文化脉络。第三章则体现了本书与理论对话的视角，涉及危机理论、社会认同理论、焦虑及不定性的管理理论以及生命历程观点等。第四章则探索十二个家庭在选择入住机构前，两个世代寻求资源的经验。第五章检视世代之间面临家庭照顾危机，启动入住机构决策过程中微妙的权力关系。第六章分析成人子女如何协助他们的父母处理对养老院照顾与生活调适的不安全感；第七章探索中国婴儿潮世代如何思考他们未来长期照顾的需求；第八章则探讨本研究对于理论与政策方面的启发。

面向国外读者以英语书写，陈琳不只勾勒出上海老年照顾模式从家庭照顾转向养老院照顾的流变脉络，同时也帮助读者聚焦中国婴儿潮世代与其父母在面对老龄化社会所产生的养老安排问题。

她首先从社会人口转型的历史脉络说明养老模式的发展与国家力量的关系：在中国经济改革、实施一胎化政策控管人口之前，建国发展阶段的毛泽东时代，政府将人口视为对抗资本主义、促使国家强大的力量，因此改变过去统治者强调“以人口控制确保国家未来人口资源可利用性”的治理价值观（p.2）。中国的婴儿潮世代即是大量出生于20世纪50年代到60年代的族群，亦即今日中国社会人口众多的中年与老年族群。随后邓小平启动了中国的经济改革，以一胎化政策严格控管家庭生育，人口出生率因而降低。但同时，随着中国现代化、都市化的发展，城乡差距更为明显，人口聚集于都市的比例大幅增长，加上生育率的下降，老年长期照顾的需求也随之增加。

对于今日社会老年人仍然倚重家庭照顾，并且由家庭负担父母养老的主要开销，作者认为，正式长期照顾资源的稀缺不只因为政策规划上缺乏前瞻性，也不只出于对中国传统文化价值观的坚持，还反映出家庭对于将父母送入养老院的抗拒。此外，陈琳指出，国人对于机构养老接受度低也与国内对于“三无”农民与城市贫民窟的联想及污名化有关。

在居家养老的选项中，虽然雇用小时工或保姆协助照顾老人是多数都市区家庭照顾者可以采用的方法，但一般家庭对于雇用人力的品质有疑虑；而社区老年长期照顾以及政府资助的老人日托中心的建设才刚刚起步；医疗机构的老年医学也才萌芽，即使是在大城市，整体老年长期照顾医疗系统与网络的建置工程仍十分缓慢。在这样的脉络之下，养老院机构养老成为一种新兴的选择。

作者选择上海一家典型的由政府补助的非营利养老院作为调研场所。该养老院居住空间类似医院病房设施，三人共享一间房，生活与社交活动模式十分规律。该院有120名员工，包括20名有证照的医护人员，而入住长者依照其健康与行动能力分成三种类型，共有320位，平均年龄为82.3岁。这样的养老院代表着大部分居住于大城市、有长期照顾需求的长辈及子女对于理想养老机构的选择：“收费低廉、政府辅助的设施以及由政府训练监督的员工”（p.16）。尤其，作者在第二章从政策面与社会文化面勾勒出中国都市养老院发展脉络的全貌，以此凸显出机构养老的照顾模式为何对父母与子女而言都是困难的决策与选择。

如同过去许多研究已经指出的，在华人社会传统中，亲子互惠关系是孝道的基础：父母抚养孩子长大，孩子长大后要照顾父母。因此华人长者在传统上比西方社会更依赖家庭照顾。也因为“尽孝道”在中国社会中仍然是影响子女实践社会角色的核心概念，中国都市化的发展对于家庭义务的表现造成了直接冲击与改变。另一方面，地方政府也将养老的责任义务更多地置于家庭内部，例如上海政府在 2012 年提出了对高龄照顾的“90+7+3 计划”—— 90% 的长者需要居家照顾，7% 会仰赖小区日照中心，而 3% 会住进养老中心。政策期望老年长期照顾的需求在个人家庭中解决，而非借由在地社群资源来满足。除此之外，有别于西方社会个人与家庭会在其他长期照顾资源仍然充分的情况下便将入住机构养老作为选项，亚洲传统家庭时常是等到家庭所有长期照顾相关资源穷尽或超出家庭负荷、产生危机时，才将机构照顾列入老年安置与居住的选择。

陈琳将家庭照顾决策的过程拆解出四个关键视角，来理解上海养老院住户与其子女从居家养老到机构养老的心理调适与赋予意义的过程：首先是启动考虑入住机构的决策过程，借助危机理论加以讨论；第二个关键视角则是决策过程中代际沟通对入住机构决策的影响；第三是以未知管理的理论来理解不同世代如何思考与面对入住机构带来的未知情况；第四则从生命历程的角度来发现照顾的决策过程在两代人的生命过程中所产生的不同意涵。作者接着透过父母、子女两个世代对于入住决策过程的访谈资料，呈现并分析了入住机构之前家庭采取共居的方式照顾父母、照顾父母决策与实际执行的性别角色，以及卫生保健政策的重新建构所产生的人际关系变化等面向。

在第八章结论中，陈琳进一步聚焦两代对于机构照顾的看法与理解如何在决策过程中产生改变，并且分析影响两代观点改变的因素。她发现在入住养老院的决策过程中，长辈对于子女尽孝的实践与期待更加在意的是家庭情感关系的维系，因此可能一方面期待子女尽孝回报父母的养育之恩，一方面仍会保留部分有关长期照顾方式的决定权力。

而参与父母入住机构决策过程的婴儿潮子女，同时也是一胎化政策下的父母，对于孝道的实践的态度是相对矛盾复杂的。找寻合适的养老院是他们因应现代化、都市化生活，在尝试家庭共居照顾模式后“尽孝”的新模式。而他们在面对自身未来老年安养规划时，不仅倾向将养老院作为其长期照顾的主要选项，同时也了解老后生活无法单靠子女“尽孝道”的家庭照顾来实现。

本书一方面透过老年照顾者与被照顾者的两代决策历程与价值观对话，反映出中国社会老龄化对于家庭与亲属关系实践的影响，同时也为社会工作与政策规划领域提供了丰富的第一手资料。在结论之后，作者将研究方法的发展与分析工具列在附录中，来帮助读者了解调研设计与方式。然而，比较可惜的部分也在于，本书在结构安排上，将中国社会老龄化的政治经济与社会文化脉络、入住养老机构决策过程的理论架构、叙说为主的访谈内容以三个块状形式呈现。在决策过程的模式归纳分析中，呈现出由理论架构映照出的复杂现实状况，但难以借由访谈对象的经验叙事，看到个别家庭两代成员访谈所交织出互为主体的家庭文化价值观样貌。同时，研究设计的方法论、研究者田野关系的探讨、研究细节与流程设计统整为“附录”，一方面易造成将研究设计与方法视为操作工具而非分析视角之误解，另一方面也缺乏有关理论对话与研究方法论之间的链接。

尽管如此，本书从社会工作研究角度出发，关注孝道实践的代间沟通过程，让读者对中国老龄化社会的老年长期照顾议题，特别是有关家庭孝道价值观变迁的时间与空间现实脉络，有更厚实丰富的理解。同时也希望能够借此引发更多中国老龄化社会的研究，促进家庭代际沟通与价值观的对话与相互理解，以帮助国家政策规划者、社会工作实践者与家庭照顾者针对老后人生的照顾需求，发展出具有多元文化意涵的养老模式。

评《中国西南的拉祜族：对边疆少数民族边缘化的反应》

王瑞静*

The Lahu Minority in Southwest China: A Response to Ethnic Marginalization on the Frontier. Jianxiong Ma. London and New York: Routledge. 2013.

《中国西南的拉祜族：对边疆少数民族边缘化的反应》，马健雄著，劳特利奇出版社：2013 年。

中国西南地区的少数民族一直是人类学研究的宠儿，而其中又以藏族、回族、白族、纳西族和傣族等最受人类学家关注。在《中国西南的拉祜族：对边疆少数民族边缘化的反应》一书中，马健雄博士选择避开上述那些有名的少数民族，而去讨论像拉祜族这样相对而言规模较小的少数民族的生存现状。该书以对云南省澜沧拉祜族自治县黑河谷拉祜族聚居地长达十五年的田野调查材料和大量地方历史文献为基础，既细致呈现了当地拉祜族内部的宗教信仰和社会组织，又极具

* 王瑞静，重庆大学人文社科高等研究院。

历史深度地讨论了近现代以来当地拉祜族与汉族进行社会经济政治互动的动态过程。该研究表明，不同历史情境中的国家是形塑少数民族社会文化生活及其变化的最重要力量。其创造的汉族与少数民族的身份二元论（identity dualism）将拉祜族边缘化，使之因为自己的族群身份而倍受歧视和苦痛，从而想努力逃离。这个逃离过程被马健雄博士称之为重置（resettlement），包括妇女外嫁给汉人，男人酗酒，甚而自杀。这就是当地拉祜族面对近三十年来的社会巨变做出的社会文化反应。

自 18 世纪末，拉祜的族群身份在该群体与清王朝和汉族移民的冲突过程中建立起来。围绕着拉祜厄沙天神的宗教信仰，一些汉族和尚发展出“五佛五经”制度，带领拉祜进行抗争。这些宗教运动同时也是政治运动，是为了对抗开疆拓土的清王朝以及汹涌而至的汉族移民（具体过程参见马健雄的另一部著作《再造的祖先：西南边疆的族群动员与拉祜族的历史建构》）。因此，对清王朝、中华民国以及其后的中华人民共和国而言，拉祜的宗教信仰及身份政治系统是一种反国家的力量，必须将它镇压乃至摧毁。而对拉祜而言，不管是哪一种政权，国家代表的总是汉族的利益，国家和汉族是等同的。换言之，拉祜族与汉族的身份二元论是在西南边疆化的过程中建立起来的。汉族处于强势地位，而拉祜族一直处于弱势地位，是被欺诈和盘剥的对象。这一关系模型在拉祜族的经济生活和历史经验中被一直维持，被一再确认，最终成为一种主导的政治框架，影响力延伸至今。

时至今日，对黑河谷的拉祜村民而言，那段远去的历史还依旧嵌在他们的宗教信仰当中。在其信仰当中，厄沙天神创造了两个世界：一是生者世界，一是死者世界，人的灵魂在生死两界轮回。这种生命轮回之说使村民得以平和地面对死亡。而来自死者世界的两种灵怪，ne 和 do，又会给村民带来疾病和死亡。人的祖先控制着 ne，一旦有人行为不端（wrong behaviour），祖先便会驱使ne来咬噬犯错者，使其生病，而这些病必须通过叫魂仪式才能够治愈。而一村之中必然会有些家庭饲养着do，对这些家庭的冒犯往往会丧命，因为他们会放出do来报复，而do的咬噬会致人死亡。人们对这两种灵怪总是充满恐惧，而对自己又是自我否定。因此，在日常生活当中，他们只能通过规范自己的行为来降低被咬噬的可能。值得注意的是，这些信仰并不是所有拉祜族共享的文化特征。通过对临沧拉祜族的调查发现，该地拉祜族并没有厄沙信仰，也没有生死两界之分及其它相关信仰。这是因为他们没有像澜沧拉祜族那样，在历史上经历了千禧年运动。换言之，黑河谷拉祜族的宗教信仰与其历史记忆息息相关。他们对来自“另一个世界”的威胁的恐惧与其在历史中遭遇到的来自他者的压力联系在一起。面对这些外部威胁，他们同样无力抗衡，只能小心翼翼地规范自己的行为，或者逃离现世。

历史上的五佛五经运动，不仅创造了拉祜族的宗教信仰体系，也确立了他们的社区权威体系。黑河谷拉祜族建立在一个以双边、非线性、非等级亲属制度为基础的社会制度之上。该社会以性别平等为准则，社区生活也是一个由地位平等的夫妻构成的亲属网络。孩子，不分男女，对父母的田地都有平等的继承权，这也成为父母与孩子之间最重要的关系。因为生命轮回在生死两界，这个亲属体系又是非线性的。因此，当地拉祜族的社区缺乏一种建立在亲属制度之上的凝聚集体的内部整合机制。对五佛五经运动的研究则表明，与厄沙天神信仰相连的宗教力量才是其社区权威的来源。换言之，在历史上当拉祜族被动员起来反抗国家时，社区权威源自宗教信仰而非亲属制度。厄沙信仰在拉祜族与清王朝对抗的过程中被摧毁，只给村民们留下了“厄沙逃离”的历史记忆以及“厄沙会回来”的希望。不过，在此基础上的社区权威则一直保留到了澜沧拉祜族自治县成立，被地方干部所取代。相比之下，亲属制度因为与社区权威相分离，在权威被摧毁或者取代之后，仍然能够在日常合作和农业生产中运行良好。不过总体而言，社区权威被毁使得拉祜族的内部防御机制彻底坍塌，只能接受被一再边缘化的命运。

作为新的社区权威，地方干部并不能为拉祜村民谋利益。虽然有些地方干部来自拉祜群体，然而他们自诩为“拥有汉族思想的拉祜”，并且在汉族中心主义以及社会进化论的框架下一再将拉祜族贬损为弱小、愚昧、落后的群体。他们把持着扶贫资源，却因腐败无能不能使拉祜民众受益，反而把责任归咎于拉祜的“落后”，以期掩盖自己的管理缺陷。在教育方面，乡村老师也一再重复着类似的话语，他们无视当地少数民族地区对全国标准化教育的不适应，只是简单地把教育的失败归咎于拉祜族的愚蠢和落后。这些话语在扶贫项目和教育中一再重复，逐渐被拉祜族内化，使之感到作为一个拉祜是一种耻辱，从而自轻自贱，自我否定。而这其中体现的“先进的汉族”与“落后的拉祜”的对比，实际上就是19世纪以来形成的“汉-拉祜”政治框架的延续。与此同时，由于缺乏受过教育的合格的地方精英参与政治生活，本已处于弱势地位的拉祜族被愈发边缘化了。

拉祜族集体性坍塌的一大社会后果是，妇女大量外流，嫁给汉人。这其中有妇女的个人意愿，但更重要的是地方干部与婚姻中介的合作与共谋。他们再次运用“汉-拉祜”的关系话语，宣称拉祜妇女外嫁是好事，因为拉祜贫穷落后而汉族是现代先进的。另外，基于对拉祜文化的误解，地方干部指责其近亲结婚导致体力智力低下，从而使妇女外嫁合理化，也使拉祜多了一重污名。对拉祜妇女而言，既然在如今现代化和市场化的环境下，拉祜身份成为一种文化负担，外嫁汉人确实成为逃离这个负担的出口。不过，妇女外嫁汉人也意味着，倡导性别平等的拉祜文化败给了男性主导的汉族价值，也无疑是对拉祜族性别平等理想的贬损。

拉祜族集体性坍塌的极端社会后果是自杀率攀升，尤其是男子的自杀率。自杀并不是拉祜族的文化特性，而是自2000年以来各种社会经济政治因素多方作用的结果。疾病、破产、酗酒、食物短缺、家庭纠纷、妇女外流是自杀的主要诱因。但这后面更深层的社会原因在于，国家在当地渗透的过程当中摧毁并取代了他们的社区权威，然而地方干部控制地方资源，使得拉祜族被排除在经济与政治资源之外。而关于“汉-拉祜”的关系话语也从精神层面摧毁了作为一个拉祜族的自尊。面对汉族/国家带来的压力，拉祜族无力自保，只能逃离到另外一个世界，这是另外一种“重置”。如前所述，生命轮回在生死两界，而两界都是一样的。一旦无法忍受活人世界，死人世界也就自然成为逃离现世苦痛的出口。无论是哪一种逃离，外嫁、酗酒或是自杀，都是拉祜族追求个体主体性的努力，但也恰恰显示了拉祜族集体主体性的垮塌。这就是一个弱小少数民族在当今现代化和市场化环境下的生存现状。

在本书中，马健雄博士的最大贡献在于，创造性地将少数民族的民族志和西南边疆化的历史结合在一起，不仅揭示了地方少数民族文化逻辑形成的历史情境和过程，也再现了在地方民族日常生活中的历史实践，以及历史记忆对当下地方权威和民族互动的影响。历史不再是一块装饰性的背景幕布，而是确确实实与地方文化社会生活紧密相连的记忆和实践。而通过对史料的挖掘和运用，有关地方宗教和社会文化的民族志也不再是静态固化的，而是呈现出其在西南边疆化背景下发生、发展和变化的动态过程。毫无疑问，该书给有志于西南研究的人类学者提供了一种新的研究路径。

编后记

几经延宕和周折，《中国人类学》第二辑终于要付梓出版了。提笔写此后记之时，新冠疫情正在肆虐全球。而本期是医学人类学专辑，时间的巧合，愈发彰显出了主题的重要。

医学人类学是当今人类学蓬勃发展的一个分支，着重探讨不同文化和社会传统中对生命、健康、疾病的医学认知体系及对病患的医疗、防护及管控的理论和方法。人类文化多样性在医学领域也表现得丰富多彩。正如本专辑特约编辑赖立里和冯珠娣在前言中指出的："医学人类学并非专攻一门的学问，而是一个多元化的领域。我辈学者在探究医疗的社会史、健康与疾病的诸面相时，殊途同归地认识到，历史上存在着远远不止一种'医学'。"世界各国对新冠疫情的不同反应、认知、理解和措施，就如一面万花镜，将各种"医学"都推向前台。从医学人类学的角度考察审视本次新冠疫情，将会催生大批新的研究，加深人类对全球化时代疫情的理解，故冀望本专辑对未来研究能起到抛砖引玉、参照比较之效。

本专辑共收录五篇医学人类学论文和一篇对冯珠娣教授的访谈。具体内容，两位特约编辑在前言中已有详细介绍，此不赘述。本期的【理论与实践】栏目，收录了尹绍亭教授"生态博物馆与博物馆人类学：回溯与反思"一文，详细介绍了苏东海先生在中国倡导生态博物馆的历史，读来尤为亲切，因为我与苏先生相识已近20年。2001年夏，我从威斯康星大学回国做调查，和苏先生相遇北京，一见如故。我深为苏先生的理念和热情所感染，就和他一同赴内蒙达茂旗考察建馆地址。苏先生当时虽已年逾古稀，但精神矍铄，童心未泯，在大草原的金色夕照下，和蒙古族大汉比试摔跤（仪式性的）的身影我至今历历在目，深刻脑海。我后来又南下贵州，走访最早的几家博物馆，参与2005年贵州生态博物馆论坛。虽然不是自己的主要研究方向，但这些年来我一直关注生态博物馆的发展和研究。往来北京，常常会去苏先生前门大街寓所探望欢叙。本期发表的生态博物馆一文，表达了学术界对这位鲐背老人及中国生态博物馆之父的敬意。

最后，请允许我对两位特约编辑及论文和书评作者表达由衷的感谢！谢谢你们的理解，耐心和宽容。由于种种因素，编辑工作遇到的难度超乎预料。如果因为出版的延宕给各位带来不便，请接收我诚挚的歉意。短短数月间，席卷全球的疫情已经深刻地改变了我们的世界，希望这本迟到的医学人类学专辑，因缘际会之下，能发挥更大的学术参照作用，以不负所有编者作者的努力。

周永明

二零二零年仲春 于威斯康星麦迪逊